Praise for *Landfill*

"*Landfill* is an important and entirely brilliant book. It's a love letter to gulls and their charged relationship with humans, but it's also a deep meditation on difficulty and waste, on the beauty of the disregarded, and on what we make of matter out of place. There's love and death here, fear, fascination, hope, and the breaking of the world. Dee has written an absolute triumph."

—**HELEN MACDONALD**, author of *H Is for Hawk*

"The British literature of birds that includes Gilbert White with his swifts and martins, Helen Macdonald with her hawks, is a rich one. But Tim Dee's own obsession with gulls also leads him to contemplate the landfills over which they often preside and the alarming changes to our landscapes with which they are associated. His alertness to factors in ecological health results not in a jeremiad, however, but instead in an exploration of surprising parallels between evolution in the biological realm and the slow siftings of memory and culture. *Landfill* is a remarkably venturesome, robustly voiced, and illuminating book."

—**JOHN ELDER**,
author of *Reading the Mountains of Home*

"Like coyotes, ravens, raccoons, and other resourceful urban wildlife, gulls frequently face our scorn, and sometimes our bullets. In his delightful jaunt through gull taxonomy, behavior, and lore, Tim Dee casts his feathered protagonists as indomitable heroes of the Anthropocene—thriving in our cities, colonizing our culture, and repurposing our trash as treasure. Next time a gull snatches your fries, you'll find yourself not cursing a petty thief, but admiring one of our planet's grittiest, savviest survivors."

—**BEN GOLDFARB**, author of *Eager*

"Tim Dee's restive and gorgeous prose pulls readers into the wilds of the modern urban landscape, where gulls and humans wander side-by-side with ancient poets, Victorian novelists, and Madagascar nighthawks. In this small book Dee asks—and beautifully begins to answer—one of the largest questions of our time: How do we live with attentive grace and wisdom alongside the varied coinhabitants of our imperiled, complex, and beloved earth?"

—**LYANDA LYNN HAUPT**,
author of *Mozart's Starling* and *Crow Planet*

"Evolution is fluid, and the urban gulls of Tim Dee's *Landfill* embody this ever-changing world in action. Tracking difficult-to-categorize gulls and the people who know their habits best, Dee alerts us to the heavy-laden meanings we lay on the wings of others, even as he revels the ways in which gulls continue to fly beyond our grasp. Familiarity need not breed contempt. As Dee shows, it can breed fascination."

—**GAVIN VAN HORN**, author of *The Way of Coyote*

LANDFILL

*Notes on Gull Watching and Trash Picking
in the Anthropocene*

TIM DEE

CHELSEA GREEN PUBLISHING
White River Junction, VT
London, UK

Originally published in the United Kingdom by Little Toller Books in 2018 as *Landfill*.

Printed in Canada.
First printing February, 2019.
10 9 8 7 6 5 4 3 2 1   19 20 21 22 23

Our Commitment to Green Publishing
Chelsea Green sees publishing as a tool for cultural change and ecological stewardship. We strive to align our book manufacturing practices with our editorial mission and to reduce the impact of our business enterprise in the environment. We print our books and catalogs on chlorine-free recycled paper, using vegetable-based inks whenever possible. This book may cost slightly more because it was printed on paper that contains recycled fiber, and we hope you'll agree that it's worth it. Chelsea Green is a member of the Green Press Initiative (www.greenpressinitiative.org), a nonprofit coalition of publishers, manufacturers, and authors working to protect the world's endangered forests and conserve natural resources. *Landfill* was printed on paper supplied by Marquis that contains 100% postconsumer recycled fiber.

ISBN 978-1-60358-909-3 (hardcover)—ISBN 978-1-60358-910-9 (ebook) —ISBN 978-1-60358-911-6 (audiobook)

Library of Congress Cataloging-in-Publication Data is available upon request.

Chelsea Green Publishing
85 North Main Street, Suite 120
White River Junction, VT 05001
(802) 295-6300
www.chelseagreen.com

RECYCLED
Paper made from recycled material
FSC® C103567
www.fsc.org

In history, as in nature, decay is the laboratory of life
KARL MARX, *Capital*

'Whatever's it all about, John?'
'Whenever is it all going to end?'
LAVINIA GREENLAW, 'Blackwater'

this is for Jenny Dee, Tessa Hadley and Tom Nichols
dumped, with me, on the family line

Contents

Checklist

I tried to give up birdwatching in my early twenties. The scene of my abandonment was a gull roost on a reservoir. From a hide on the banks of Chew Valley Lake, outside Bristol, I often watched gulls coming to land on the water to find a place to sleep. Many common gulls did this and occasionally one or more ring-billed gulls, vagrants from North America, who had got caught up on the wrong side of the Atlantic, came to rest with the regular crowd.

In the hide, if others were with me, there was a chance of picking out a ring-billed gull. But on my own I despaired and never managed anything. Thinking of what I was despairing about despaired me further. My wanting to see a ring-billed gull was stopping me seeing – the shakedown of thousands of birds arriving out of the dying light, and then the exquisite distribution of these white forms across the darkening water. I felt stupid to have misread the truth of the scene. I put away my binoculars for much of the next three years.

Before I quit I had seen most of the gulls then known in Britain. The gulls I grew up with were herring, lesser black-backed, great black-backed, black-headed, common and kittiwake. Along with ring-billed, the gulls I began as a teenager to look for and chase as rare birds were:

little, glaucous, Iceland, Mediterranean, Sabine's, laughing Bonaparte's, Franklin's, Ross's and ivory. Some are Arctic species; some are American.

In the recent past, revisions to our taxonomic understanding have released more gulls that might now be seen in Britain. Most notable among these are yellow-legged and Caspian gulls. There are also smaller numbers of other split species and assorted subspecies, races and hybrids, including American herring, Thayer's, Baltic, Azorean, Viking, Nelson, and more with names still stuck in Latin. All are hard to see and few are easy to tell apart. Yellow-legged and Caspian gulls are the species whose observed presence in Britain has prompted new interest among birdwatchers in the family as a whole. They are also gulls that have been watched extensively on landfill sites. They are the birds that started this book.

The first rare bird of any kind that I found was a Mediterranean gull at Oxwich in the Gower. I was thirteen. My friend Richard was with me, and we wrote up a description of what we saw and had our bird accepted by the county recorder. Richard is dead now. The last time we spoke by phone, a few years ago, he was in the back of a car taking him to a business meeting. Forty years on, the gull was mentioned.

After the Mediterranean gull, I got serious and twitched other *rares*: an ivory gull at Chesil Beach in January 1980, an astonishingly polar-white bird; and before that two Sabine's gulls at Sheringham on a wild autumn sea-watch; one laughing gull at Yarmouth (or was that a Franklin's gull?); a Bonaparte's somewhere, perhaps in Cornwall. I should check my notebooks, except I wasn't a good diarist then, especially not on the chase, when everything in front of me was too exciting to be captured by standing to one side of it and writing it down. Besides, I was intimidated by my fellow birdwatchers: the beautiful lifelike sketches of Laurel Tucker, who was often in the front seat of the car we shared from Bristol to get to the target bird, and who is dead now too; or the meticulous counts and annotations of Antony Merritt, who wrote in pencil and then sprayed his pages with fixative to preserve his lists, and who now is sick and has been, for decades, cruelly confined indoors.

It was a gull that also got me back into birdwatching.

I worked for a time in bird conservation after I left university. For three years birds were my day job, and usually that left me wanting to do other things when I wasn't at work. But the lure of some species never dims. An adult Ross's gull – a remarkably pink bird – arrived in Norfolk in May, 1984, and my boss, Nigel Collar (more from him later), was similarly surprised by his resurgent appetite; he drove us to collect the prize at Titchwell where it sat like a melting raspberry-ripple ice cream on a muddy island.

★

Only in the past decade or so has a subgroup of birdwatchers interested in gulls been visible. Some say they are going gulling; others call themselves larophiles (*Laridae* is the Latin name for the gull family; *Larus* for the genus that includes the large gulls). When I spotted this new interest I wondered how and why some of my gang had fallen in so deeply with one family of birds not much regarded by other birdwatchers and often derided by non-birding people. It was striking, too, that most often this new attention was being paid in the towns and cities where most of us live, and the dumps where we discard the leftovers of our living. There are two reasons, it seems, why gulls have loomed as large as they have: first, simply, gulls were newly nearer and more visible to most birdwatchers than other birds; and – more complicatedly – taxonomic changes had magicked more birds for us to see.

This book tries to unpick this story: how, over the past one hundred years, gulls have made their way among us in a man-made world and how, more recently, various people have met them there. But it is not a book of natural history or of anthropology. My real interest is in exploring what these facts of gull-life and gulling-life have done to our minds. To do this I have watched the birds themselves in the company of people who know a lot about them, but I have also read poems, plays and novels, and seen films and grappled with philosophical essays that have all found reasons to be concerned with gulls.

Studying both organising (which could include bird systematics and the recycling of rubbish) and disorganising (anarchic bird lists and chaotic dumps), and then the organisation of the disorganised (revisions to avian taxonomies and to the categories of rubbish), took me to thoughts that weren't about gulls, or even birds, and there are pages here that have nothing flying through them. Forgive me for that, but trust me that it was gulls that led me there.

All this that follows then, is to try to discover what I think of these birds, and how all of our thinking has shaped the way we know them; all this, to ask how changing the way we consider an animal might alter its meaning.

*

Nowadays gulls are trash birds, the *subnatural* inhabitants of *drosscapes*. Their coming among us has lowered their sea-bird status. Today they are seen as déclassé and mongrelising in their habits. Some have also been demoted from whatever former taxonomic security they had. All have become in-between birds in an in-between world. By moving towards us, they've risked becoming like us. Bin chickens, some call them.

At first, I thought writing about this might describe an impoverished experience: birders turning to gulls because they are the only birds around. I also thought watching the watchers and the watched might be melancholic, or darkly funny: men leaving their homes and their families to spend time peering at arsey birds in some of the arseholes of the world. There is a bit of this, but it turns out more substantially that the meeting of gulls and people is exuberant. It has an excess that could be called joyous.

As the birds worked our utility places and our waste, I saw that birders were processing the gulls, picking through them, finding new things to know and to understand, finding value in creatures otherwise labelled shoddy or dreck. That means, though the gullers don't often admit it, new things to love. This enthusiastic organising of life (an action potentially lovable itself) in the midst of the organising of what we could call death (the sorting and rendering of our waste) is gripping. Landfill means more than just a tip for the end of things. It is also a description of how we have worked the rest of the living world, learned about it, named and catalogued it, and have thus occupied or planted our planet, filling the land.

White Goods

In Greek and Roman mythology, one of the entrances to the underworld was a sulphurous lake called Aornos or Avernus. It is near Naples. The name, meaning birdless in Greek, might describe the effect of toxic fumes, rising from the poisoned waters, which were supposed to kill overflying life. Modern Naples has had decades of garbage disposal problems. The criminal Camorra had more than a hand in waste management, and the region has been shipping its junk by boat to be burned in the Netherlands. I've watched ships offshore at Naples marked by towers of gulls seeking the rubbish. The world works, or doesn't, like this. Aornos sounds like a landfill death-pit to me and, doubtless, there will be gulls there today, toxic or not. The world also works like this.

'A certain desolate area of land.' F. Scott Fitzgerald knew the early landfills of New York State and put one in *The Great Gatsby*: 'a valley of ashes – a fantastic farm where ashes grow like wheat into ridges and hills and grotesque gardens where ashes take the form of houses and chimneys and rising smoke and finally, with a transcendent effort, of men who move dimly and already crumbling through the powdery air.'

Fresh Kills on Staten Island in New York State might have auditioned for Fitzgerald. It sounds like a perfect name for a dump, but the place doesn't mean what it is called. 'Kill' is a Dutch word for a stream. The water there was once good. And now the landfill site, which previously served New York City and became the largest man-made structure in the world, having been reopened to take human remains and building waste of the 9/11 attacks, is to be plugged and capped and metamorphosed over time into an urban park. The same has happened at the equally well-named Mucking in Essex (a fabled gull spot).

★

'The waste remains', said William Empson in his poem 'Missing Dates', first published in 1937. The phrase becomes an infectious refrain that slides through the villanelle: 'the waste remains and kills'. Things fall apart and their ruin is toxic, 'Slowly the poison the whole bloodstream fills'.

On the dumps I've been to, the first half of Empson's assertion is manifest. There are accumulating metres of remaining waste: 'the slag hills / Usurp the soil,' the poem says; that is true at places like Pitsea in Essex. But the second half of Empson's phrase is less demonstrable. Plastic junk lives on, horribly, and toxins persist on dumps, but they are also – strangely – *lively*.

At the point we identify anything as waste, even though up until then it has been ours, we don't want it and we don't like it. Anything can become waste or dirt in this way. Dirt, in Mary Douglas's memorable formulation (developing William James's thoughts) in *Purity and Danger*, is simply matter out of place. And therefore places designated for dirt stir complicated emotions. What we had made our own is now dirty, and we want it away and gone. But the waste doesn't necessarily kill where it is, in situ it can live, and even nourish and sustain. It is like this for the gulls at landfills. It is also like this for all sorts of human activity too. We are a waste-making species like no other, but we are also workers of waste, recyclers, ragpickers, archivists, librarians, archaeologists, historians, bricoleurs and gullers.

★

There has been a gull moment and it is coming to an end. Populations boomed through the throwaway decades of the 1960s and 1970s. But now the land has been filled. We are managing the end of things and assorted half-lives. The dumps are being grassed over and converted into country parks. Today, the recycling or incineration of food waste is commonplace, and edible trash is rare at tips.

At Pitsea landfill, the gull-men negotiate with the waste trucks, holding back a cart with restaurant leftovers until they have readied their ringing nets. Much else continues to arrive, but there is little to eat. Gull numbers are falling. There are salmon and sea horses in the Thames near Pitsea, and that is surely good. The waste barges with their headaches of hungry gulls no longer float down an open sewer to Essex. We desire a cleaner world. Meanwhile, the gullers are making a list of what they know they are losing.

★

Some people have got more interested in watching gulls in the recent past because gulls seem to have become more interesting. The herring gull that we once took to be one species living right around the Northern Hemisphere is now known as half-a-dozen related but separate species. New ways to plumb the birds' DNA told us this. Previously, these birds were lumped together as lower taxa – forms, races or subspecies of a single species. But now they have been split. The gulls haven't changed – or rather they haven't stopped changing, as natural selection continues to operate on all of them, as on us. But we have found new ways to get closer to what might be their truth. Having decided that several deserve a new life-status, we have told ourselves a new story about them, given them new names, and – if we are thus-minded – new values, since a species is worth more than a subspecies to lots of birders.

Before there were declared gullers or larophiles, a few observers had noticed that various incarnations of herring gulls occurred in Britain, some commonly, others rarely, some breeding, others visiting; but since they were all termed herring, and their sub-specific differences were hard to tell apart, most birdwatchers left them alone. Knowing them as herring gulls was enough.

I grew up as one of those birdwatchers. I learned herring gulls, I learned lesser and great black-backed gulls. These three looked both different enough to be separate species, and close enough to their neighbours to be recognisably in the same large gull family. Adult birds were easy to tell apart when they had to be allocated to just one of three species: herrings were the grey-backed ones and the other two were darker, and one of those was a giant. The young were always more of a challenge. The messy and overlapping progression of browns and greys, like various muds and skies, through four or more years of immature plumage, meant that I left them undifferentiated. I was lax, but it wasn't a heinous crime. There were only three species. And none of them were rare. Or so I thought.

Over the past twenty years, after the splitting, when new names were given to new species, birders wanted to be able to identify the birds in the field for themselves. Suddenly, as it seemed, up to a dozen new gulls might be passing through Britain. DNA was indicating that the birds were species apart, but could they be separated without a blood test? Did they look or sound or live differently? Could close attention allow you to be certain? So the guller was born: dump devotee, attendant at sewage outflows, negotiator of access to assorted badlands, sifter of species in the dinge of waste places, scrutineer of remiges and rectrices, patient teller of moons and spoons, windows and pearls.

Five years ago I had not knowingly sighted in Britain a yellow-legged or a Caspian gull, the two major additions that the splitting and shuffling of the gulls has delivered to the British list of birds. And I got interested because I couldn't see them. These were the first birds in my familiar world that I was blind to, the first that had bewildered me and sent me packing without a name in my notebook. It wasn't their fault, but I felt lost among friends, as I had not previously been in Britain. I was never a great birder but, since I began as a child, I have not been as bad a birder as the new gulls were suggesting I was.

★

In my lifetime gulls have come towards us. Most other birds have gone in the opposite direction, but the gulls have bucked the trend. In part we made them do so; in part the birds elected to fly that way. And they continue to tell how something of the once-wild can share our present world. Calling them *seagulls* is wrong – that was one of the first things I learned as a novice bird-boy. They are as much inland among us as they are far out over the waves. Yet, in fact, this state of life for them is new. Over the past hundred years, human modernity has brought gulls ashore. They have lived in our slipstream, following trawlers, ploughs, dust-carts. In this way they are more our contemporaries than most birds. They live as we do, walking the built-up world and grabbing a bite where they can. Of course they also lay eggs and fly, but they have taken a place in the chapters of our lives as few other animals have.

At the moment, this largely disturbs us: we have started fearing gulls for getting good at being among us, and we've begun to hate them for it. We see them as scavengers, not as entrepreneurs – as aliens, not as refugees. They steal our chips and kill our chihuahuas. They are too big for the world they have entered. Some of this hatred is particular to the times, and some is a resurgent rivalrous antagonism that almost any other creature on Earth can trigger in our species – that dark loathing we can find in ourselves for any living life. Yet, even besmirched like this, the gulls keep us company. And they'll be with us for the duration of this, our late hour.

★

'To cleave' is a verb all taxonomists must fear, because it means to split and to lump, to pull apart and to bring together. It also describes what we do to our rubbish – and how we feel about gulls. We furl them under an arm, their beaks at our back, their legs facing upwards so that we might fix rings on them in order to know them. And we wish them gone, out of our lives, away from our chips. The verb 'dust' is a word that describes everything here, and all that is happening. It bifurcates in meaning and uncannily contains its opposite: to dust is both to clean and to dirty. It is what is removed and what remains.

We are all stardust; we are all frass. Spending time in rubbish places with their gulls and their people, we might feel the weave of the rag-rug of our world.

★

On 1 December 1963, just two weeks after it had first appeared from beneath the sea off southern Iceland, gulls were seen to land on the cooling volcanic waste that was growing into the new island of Surtsey. They were the first life form to set foot on the newest addition to the land surface of the Earth. Thereafter they continued to visit the bad black tooth that the eruption became. Now they have turned it, in part at least, green. I can't find a specific identity for those first-footers, but five gull species have since bred on Surtsey. Great black-backed gulls first reared young in 1974, kittiwakes in 1975, herring gulls in 1981, lesser black-backed gulls in 1983 and glaucous gulls in 1993. Most (apart from the kittiwakes) live in a busy mixed colony on the south side of the island. A count in 2003 noted 301 pairs of gulls, mostly lesser black-backeds.

All these gulls make nests, and on Surtsey they tore up the pioneer plant life in order to do so. But, as time went on, they also planted their own place by dropping all sorts on the island. A study of soil development showed that, within their colony, the gulls transfer 45–50 kg of nitrogen per hectare every year from sea to land, while the surrounding, barren areas receive only 1–2 kg as atmospheric deposits. The birds fertilise the ground: they defecate, they regurgitate, and they drop food remains that they have found elsewhere, and their empty nests compost back into the deepening soil. They are landfilling.

I had a day and a night on Surtsey in 2003 to mark the island's fortieth birthday. The raw material of the Earth, the coughed-up guts of our planet, was still much in evidence. But the place, so junior and new, was already ageing and shrinking, its foundation stone eroding soon after being set up. The sea was eating at the island's friable edges; it was half the size that it had been in 1967, when the eruption stopped.

Naked rock will seem so anywhere, but Surtsey's grizzled lava has a grey pallor just beneath its skin that looks like death. It's a cinder mountain. An ash heap. Its burning was its life and everything after feels posthumous.

Except for the gulls and the meadow they have made.

I tried to trek the length of the island. Treading on this old-new volcano, the colours below were mostly grey-black, with a few patches of urinous yellow and rust. I felt like I was walking an autopsy. My boots were shredded by the rough climb; I cut my hand on a lava snag. On the bald summit the sea-wind's harsh blow met foul air smoking from the hot cracks that riddle the rock. My head spun; my lips chapped. There were no birds. I slid and skittered towards the south.

There, in a sheltered bowl, wind baffled and out of earshot of the sea, a garden is growing. It is this that the gulls have made. A dozen lifted from their nests and *ugg*-ed at me. They had downy young hidden in the long tangle of meadow grasses and plants. It is an extraordinarily green sanctuary. The gulls' agitated grunts were the only sound I could hear and they stopped after I lay down on the soft bed the birds had made for me. I could smell the nearby chicks, the warm dust from the cracking cases of their new feathers, the cooked greens that they shat, and the ozone-iodine tang that their parents gave off close-to. The sky came blue above and the birds returning to their nests drifted silently over me. And there, in the gulls' home, I fell asleep.

Our Mutual Friends

Early morning Bristol. The bars, along the street where I live, recycle their glass empties of last night. It is the sound I most often wake to: a raucous shuck and slide, a bright and broken glissando, poured into a cavernous bin. Perhaps the city gulls are alerted by it too, for at this time, before the morning traffic builds and the pavement fills with people, herring and lesser black-backed gulls come most days and patrol the street for the leftovers of last night's fun. They fly back and forth, at the height of a double-decker bus, eyes down and silent, with something in their action both futuristic and ancient – drone ops and dinosaur stares.

Much of my life happens here, next to the bins and the gulls, in a flat on a busy shopping street. Below my room are a mobile phone repairer and an e-cigarette shop. Half of the pavement is taken up by big and locked dumpsters used by the neighbouring bars and cafes. About a decade ago the city council issued the residents of my building with recycling containers, a waste food tub, and a wheelie-bin for non-recyclable rubbish. The system worked for one week. My suite of emptied units never made it back to my front door or anywhere near it. All was lost in a melee at the roadside. There and then I gave up. Nowadays, I carry my paper waste, my glass and plastic to the old house where my children sometimes still live and where bin protocols prevail. My food scraps mess with everything else I shed at home into an old supermarket carrier bag that I then drop into a street-bin one hundred steps from my door. It is possible that my lumped and unsorted offering, joining every pedestrian's discards, and collected by a council worker with a trolley, is bait for those morning gulls. If not, I think my rubbish ends up as landfill.

★

The idea that gulls might be good animals to think with and to write about first occurred to me around twenty years ago, when I noticed how herring and lesser black-backed gulls had set up shop in the middle of Bristol. There had been large gulls on the city's waters for many years; it is only a few miles from the sort-of sea of the Severn Estuary, while the River Avon, a tidal finger, drives daily into the heart of the place. Among Bristol's sound-signatures are the birds' marine yelps as they navigate the Avon and the Feeder and the Cut and the other channels that broker the meeting of fresh and salt water. But in the years of a stretched decade around the 1980s, when I left the city (to study, and before I returned, with a young family, to work), the two gulls had started something new: breeding on rooftops across its centre. In that time, well within the lifespan of individual birds, a profound change gripped both species and urban gulls came of age.

Near the centre of Bristol, until a few years ago, there was an ice rink and music venue in a chunky building put up in the late 1960s. I went and skated there once as a teenager and hated it. A year later, next door at the Locarno, I watched Suicide support The Clash, and enjoyed the electronic

Americans more than the jangly Brits. There were no gulls at that time, as far as I recall. But the complex had a flat roof.

By the 1990s you couldn't miss the gulls. From early in the year until late summer that same roof was busy with herring and lesser black-backeds. They were at home. Standing next door in the multi-storey car park on Trenchard Street, you could easily see fifty or more birds. I began going to watch the colony during my lunch breaks. Level G – the highest in the car park – was open to the sky and offered a panorama of the city centre; it was good for gulling, and midsummer was a good time to go.

The first afternoon I went was hot and muggy; the felted roof grew sticky. A shrinking puddle of water attracted three young lesser black-backeds. They sipped at it. One still had down around its teenage face. Bum fluff, we used to call it in my Locarno days.

From the car park, I could see that almost every suitable flat roof across the city centre had breeding birds on it. There was even a herring gull between the hooves of one of the two golden unicorns prancing atop the Council House at College Green. Nearby, someone had planted a two-foot-high plastic great horned owl on the apex of a roof, to try to scare the birds off. I was pleased to see another herring gull dozing on its nest at the base of the shit-spattered decoy.

The gulls seem particularly drawn to this area, as if they divine its hidden waters, where the last reaches of the tidal river runs culverted and capped below the city streets. In fact the water is not so important to them. They have made a nutrient-rich sea out of Bristol's food waste and a marine archipelago out of its unlovely rooftops.

At the ice rink, jackhammers were demolishing something with a noise like the bombard of a shingle beach at the foot of a cliff. The hum of the rink's cooling fans added to the mix. The roof, with its black-tarred felt, was like a raised beach or the lava plug of a volcanic island. Dripping in the July sun at the back of the rink was a small mountain of dirty, cast-off ice, like a cache of sherbet or a colossal haul of cocaine or – if you are seagull-minded – like a wayward iceberg fretting at a northern shore.

The nesting gulls preferred the edge of the roof above, where a shallow trench had grown a thin scab of mud and a drift of bottles. Earlier in the year I had watched gulls flying over the city with mouthfuls of bright green moss or fresh grass. They were busy then, with repairs or the building of new nests.

The rooftop assembly looked casual and messy, a shanty town built on a sewage works, but this was a formalised space, a place of territories, rituals, and hierarchies of age and of species. There was even a dance floor.

An adult lesser black-backed gull landed near its two young with a piece of grey chicken skin flapping in its beak. One youngster grabbed it and bolted it down. Other young birds bleated hungrily; some tried their wings and made circuits of the roof; others walked about like disconsolate children trailing home after a summer day on the beach.

Flying ants crashed everywhere and loafing gulls picked at them. Between begging calls, the young birds made more practice launches, flapping their wings and jumping. Paired adults were re-forming their relationships; returning birds went in for bouts of head-flicking and kissing. Neighbours were in dispute, caterwauling above the din.

Moaning beggar babies were held off by stabs to the head by their mothers or fathers, but they were tenacious and advanced again, pecking at their parent's red bill-spot. A young bird walked backwards like a hypnotist leading its zombie parent, and eventually a meal was sicked up. A loitering carrion crow flew in, but the adult gull, broken from its spell, saw it off.

Bristol, it seems to me, has got something it deserved. The city that brought the Atlantic to Britain – slaves, sugar and tobacco – has drawn seabirds into its heart. The gulls are canny opportunists and worthy embodiments of the spirit of the place. And people hate them for it.

★

I've lived in Bristol most of my life. Watching its gulls has redrawn my map of the city, and half a day with Peter Rock changed it further still. He's been studying the urban gulls of Bristol and the Severn Estuary for decades – he taught art for a time but stopped to follow the birds. In late January 2017, I met Peter for a few hours after he had completed a morning routine at his health club. Early in the year he does his gulling in town, mostly by car. I took his passenger seat and within minutes we'd pulled up on a curb in the St Pauls district. Peter reached behind him for his binoculars. There was a fluster of gulls at a street bin. Two herring gulls flew off smartly but Peter spotted a colour ring on the left leg of a lesser black-backed gull that had landed on a lamp. He jammed his car into the side of the road and had his telescope out of the boot before I knew what was happening. Within moments he was back in the car writing the ring number in his notebook. The gull had been ringed at Pitsea landfill in Essex, he said, probably last year. Since he administers all the gull colour-ringing schemes in Britain, Peter knows how and when the number and letter sequences changed in Essex. 'That's one from Paul Roper's team.'

Peter drove and talked and we stopped when gulls appeared. He knew where and when this would happen. There was nothing random about our progress, nor was there about the presence of the birds. Where they jinked between buildings so did we; where they dowsed for dirty water and binned food, we tracked ugly and unshowy districts.

'Who the hell ever goes birding in a trading estate?' Peter asked. 'Apart from me!'

We pulled up outside industrial units, we surveyed school buildings and the rooftops of office blocks, we put a telescope on a wastewater outflow and sat next to a weir on a canal, we went back to Level G of the Trenchard Street car park. It was a journey around a city that I had thought I knew but which was shown to me afresh: a city of gulls.

In the 1980s, when Peter gained his 'A' Permit (the senior bird-ringing licence) there were about a hundred pairs of both large species of gulls in Bristol. Now there are 2,500 pairs. 'It's the same story throughout Britain,' he explained. 'The last national survey, Seabird 2000, listed 239 colonies across the country. I've been quietly building a register and I'm already on 523 sites; numbers have more than doubled since 2000.'

Roof-nesting gulls are now found along the western seaboard of Europe from the north of Norway to southern Morocco. All urban European gull populations are increasing, though they remain small in comparison with some of those in Britain and Ireland. The habit of roof-nesting in urban areas in Portugal, Peter told me, lags about twenty years behind the gulls of Britain.

'I hope we might see my oldest bird today, a lesser black-backed gull. He's twenty-eight. It'd be nice if he was in. I saw him at a landfill a couple of weeks ago, and he's still got the ring I gave him when he was five weeks old. For thirty-eight years I've ringed only nestlings. That way I know exactly how old they are and exactly where they come from. It might seem glaringly obvious, but it marks a big difference between me and, say, Paul Roper and the North Thames Gull Group, who catch full-grown birds at Pitsea landfill.'

We stopped at a waste transfer station, where recyclable household rubbish is gathered and sorted. If they're lucky, the gulls there can get a day's food in seconds. Peter pointed out a skip with the lid open.

'It's full of the waste that we put out in little brown tubs. Recycling lorries bring it here. Once upon a time there used to be crowds of birds before the council sussed out that they could put a net over the top to stop the gulls getting at it. I've seen a thousand birds right here, feeding, then having a wash and brush-up next door on the river, where they stand around and where it's very easy for me to read the rings.'

An enthusiast for his subject, though not a biologist by training, Peter was the first person in Britain to start studying urban gulls. He is seriously playful about his calling. On our day out, as on every gull day, he wore a splotchy camouflaged jacket. 'This is my seagull jacket, so when they fly over me they see I've already been done, so they'll crap all over you instead.'

We drove on. Another stop. We were on a road between the Cut, a tidal canal, and my elder son's old school. There were herring gulls on its roof.

'This is an interesting thing: every school now has its own flock of gulls. When the kids come out into the playground, they all drop food and the gulls come and hoover it all up when the bell goes. I sometimes see a hundred birds on the school roof, waiting to pick up crumbs or crisps or whatever the kids have chucked out of their lunch boxes.'

We scanned the Cut on the other side of the road; on its muddy bank was a lesser black-backed gull that Peter had ringed. Once the only way of studying the birds was to put metal rings on them, but they were small, grey and inconspicuous. Ringers then started putting on combinations of colour rings, stacking up to three on each leg. But if a bird lost just one, the sequence was gone and you couldn't pin down the specific bird. Peter got in touch with Malcolm Ogilvie at the Wildfowl Trust and discussed the option of using engraved colour rings for the large gulls, in the same way they were doing for waterbirds in their collection.

'My engraved rings – the first to be used on gulls in Europe – show a pair of letters with colours changing every year. They're very easy to find and read. This ring on this bird here, on the mud, is black showing a yellow code. I've used that colour three times: black showing yellow without punctuation, black showing yellow with a colon, and now black showing yellow with a plus symbol. Colour-ringing like this revolutionised how we look at gulls.

'And this bird is actually what I want, seeing as it's January and it's a lesser black-backed gull and, in the old days at least, it shouldn't really have been here yet. They used to migrate south after breeding. Things are happening with this species. This urge to dispense with migration and stay put is twice as common among urban gulls compared to rural gulls, and it may be that this is an adaptation to living in town, driving further differentiation between the two populations.'

We crossed over the Cut and headed towards some back-lane industrial sheds in Bedminster. There were conspicuous paired herrings gull on the rooftops.

'Everything speeds up in town. It smells of success. That's why there are so many birds here. There are advantages over rural life. There are free lunches, and it's much warmer than the surrounding countryside. The gulls can start breeding much earlier than rural birds. Even though it is January, supposedly the dead of winter, the breeding season has already started.'

We pulled in outside SB Fitness, a gym, in Whitehouse Street. SB stands for Sweat Box.

'This is the roof where one of my psychos lives. I have a nice video on my phone of him attacking me. He keeps on coming closer and closer and then you hear the crack as he actually hits the screen. The urban gull issue which, nowadays, causes all sorts of headaches the length and breadth of the country and has already cost millions of pounds, started over sixty years ago. For opportunist birds like gulls, the 1956 Clean Air Act – preventing the burning of refuse on tips – meant a food bonanza to be exploited. Adult survival and breeding success increased dramatically.

'The gulls outgrew their traditional colonies. Out in the Bristol Channel, Flat Holm and Steep Holm reached carrying capacity. It was inevitable that urban habitats – a series of warm islands with flat tops and steep cliffs – began to appeal. Better still, there were no predators. And in the beginning there was no disturbance from humans, who were reminded of their seaside holidays.

'Seabird 2000 estimated 31,000 urban pairs (herring and lesser black-backed) in Britain. But that was woefully under-counted. There may be more than 100,000 pairs overall. I've put a conservative total of 25,000 pairs for the Severn Estuary area alone. It's important to know these things when trying to talk to people who are looking for a solution to the urban gull problem. Even if you could remove all of the gulls in one colony – which is well-nigh impossible – you will not be able to keep the gulls out of town. I try to tell people that their schemes probably won't work. When I say probably, I mean definitely. They use netting, they put spikes up, the roofs of Bristol are dotted with plastic owls or patrolled by flying birds of prey, and a lot of council tax is being spent. It's a pity that some of the money wasn't spent on research. The big conversation about all this, which David Cameron suggested in the summer of 2015, never happened of course.

'The idea that gulls come into town for food is wrong. They are in town already. We're seeing the development of two mostly separate populations of large gulls in Britain: the rural and the urban. They mix in feeding areas out of the breeding season. But birds hatched in the city breed in the city. Similarly rural birds stay rural, with just slightly more coming to town than the other way around. I know just eleven or twelve gulls that were hatched on Flat Holm that are now in one of the Severn Estuary urban colonies.'

We moved again, to a trading estate near the Feeder canal, looking at the roof of a joinery business.

'My oldest bird breeds up there. They love asbestos roofs. In places they're covered in moss and they just have to rake it up for their nests. Here, the two gulls breed alongside each other. Lesser black-backed gulls in their rural state are dune-nesters, and herring gulls cliff-nesters and, to a degree, we sometimes see that rewritten here: the lesser preferring flat roofs, the herrings wedging their nests against chimney stacks. But they will share the same roof. And they do also sometimes hybridise. This is another urban trait. Hybridisation usually occurs as a consequence of range expansion when there's no-one of their own species to pair with. But it is usually bred out as range expands, and often the offspring aren't viable. That's not so here, where they breed together and produce viable, fertile, offspring. They can be buggers to identify.'

Our last stop was a return to Level G in the Trenchard Street car park. There was gull after gull on roof after roof.

'We talk about survival at a species level but it's also possible to get to know individuals. Each has its own way. The ringing helps track this. I see the usual suspects at a landfill, for example, whilst there are other birds that I know that you never see there. Yet, surely, they speak to one another; every bird in the area must know about the place, every bird must know everything there is to know in its own home range: if it didn't it would die.'

We're just beginning to get closer to understanding this. New technology is helping. Peter worked with the University of Amsterdam on fitting four herring gulls in St Ives, Cornwall, with GPS trackers in 2014 – the first time this equipment had been used on roof-nesting gulls. The devices sent details of where they were every five seconds. It was revelatory.

'Then in 2016, as part of a Bristol University project, we tagged five lesser black-backed gulls in Bristol and another seven in 2017. During the breeding season none of these birds went to sea, though it's only 12 kilometres away. They fed a lot on farmland, especially after grass is cut for silage.'

This ongoing work has made Peter a collaborator with avionics engineers. Discovering the feeding places and aerial routes of the gulls is a dividend of a project designed to study the birds' flight energetics. Working out how gulls cope in the complex airflows of the urban world, with its unstable windy atmospheres between buildings, may lead to new designs for tomorrow's drones.

I had one last question for Peter. Do you love them?

'Naaahhh, but they've been responsible for many adventures – they got me arrested in Morocco and made me friends in Portugal. But now I go to places where I try not to get to know anyone, so I can get to bed and up the next day early enough for the rings.'

We got ready to leave the car park. The roof of the building that replaced the ice rink next door is flat and has a good parapet. A herring gull landed and marched about on it as Peter packed away his telescope.

'These birds in front of us, at this time of the year, still early in the season, are about to buzz off for supper and then on to roost. From here they go to Waveney Sands at Frampton on the Severn, or Chew Valley Lake, or the Cotswold Water Park. They still want to sleep on the water, but it doesn't have to be the sea.'

Big Bird

By moving onto the rooftops of our buildings and sourcing our edible refuse at rubbish dumps, gulls – herring gulls more than any – have come closest of all birds to our present lives, and have gained unique admission into our places. They now arrive among us as *basic* things, like everyday items from a supermarket. No other largish bird is more commonly encountered. Pigeons in cities and pheasants in rural areas compete as members of a 'slum avifauna', but both those birds are far more in hock or surrendered to human terms and conditions. Gulls are still *wildlife*. The gulls' boom in the northern world has coincided with an industrialised and urbanised humanity. The family's fortunes have mapped ours, and vice versa. As we have evolved, so have they. In 2016, a herring gull was dyed orange after it fell into a vat of curry in Newport, South Wales. In 2018, herring gulls were reported drunk along the Devon and Dorset coasts.

It is different in the dictionaries and encyclopedias. There, gulls materialise as long-winged, heavy-bodied, web-footed seabirds with stout bills. They all have eleven primary flight feathers (the outermost is vestigial and barely visible) and twelve rectrices or tail feathers. They are mostly white and grey as adults, browner as immatures. They are highly social and gregarious; they breed, roost, feed and migrate in groups. They are noisy during most of their activities. The family is 'adaptable, opportunistic and omnivorous' (*Handbook of the Birds of the World*). Among seabirds, they are the generalists: the least picky at the buffet and the most sure-footed on land (shearwaters ground, gannets beach, gulls carry on regardless). This has helped them ashore in all sorts of ways. The herring gull 'spends much time sitting and standing, preferably at places commanding good all-round views; [it] walks, wades, and perches as necessary' (*Birds of the Western Palearctic*).

Adding up these generalised bodies worldwide, there are seven gull genera, fifty-two species and seventy-eight taxa. Five species are considered globally threatened (none in Europe – Audouin's gulls in Spain, recently imperilled, have been rescued, though remain 'conservation dependent' and continue to need help). No gull has gone extinct since 1600, the date taxonomists declare modern life to have begun.

Earlier, the gull fossil record is 'meagre'. They are believed to have diverged from other Charadriiformes (waders, skuas, terns, auks) as early as the Palaeocene, 66 to 56 million years ago, but the first true gulls come later, during the Miocene, 23 to 16 million years ago. The birds appear to have evolved around the northern Atlantic. Fossil remains of two species, *Larus elegans* and *Larus totanoides*, are found commonly in the Aquitanian Formation in France. Some researchers place these birds in a different, proto-gull group (the *Laricola*), but gulls, as we all might think of them today, appeared roughly then.

Most of the gull-world happens in the Northern Hemisphere; there are far more species in the north and there also appears to have been a lot more speciation here.

The herring gull is relatively junior in the family. It is known from the Pleistocene, or the last Ice Age, in Europe, and appears in more recent post-ice deposits, and then in human-made middens.

They have long been associated with rubbish.

Although we have happily plundered them over the years (the bones on middens would have been non-edible leftovers from when people ate gulls), it is hard to detect much *liking* for any of these big birds. The *Dictionary of Birds* (1985) characterised the then unsplit herring gull as a 'large, homogenous, successful group probably at the summit of another evolutionary line'. Its intention was surely sober but it is hard to escape a whiff of disapproval, a suggestion in this description that what has arrived is something of an arriviste – that something like a descent has taken place, or that a slum avifauna might be a slumming avifauna.

★

Gull taxonomy is messy.

The classification of the herring gull, named *Larus argentatus* – the silver gull – by Erik Pontoppidan as far back as 1763, has proved to be enduringly complex and contentious. Even now we might not have it right. All the weighty books feel obliged to describe the chaos. Scholars, otherwise impartial, cannot hide their exasperation. This is how the *Handbook of the Birds of the World* takes a breath before diving in:

> probably no group of birds has done more to challenge the biological species concept than the Herring Gull and its relatives … there is no entirely satisfactory and definitive relationship that can describe their mosaic evolution.

Gulls are evolutionarily young in the north of the Northern Hemisphere, where the ice of the last glaciation is still relatively memorable to, or operative on, many life forms. Rising up, like the land recovering after its frozen smothering, the gulls are still busily becoming themselves. The most recent common ancestry for gulls, in Europe, seems to have been from two groups of birds that re-colonised different areas after the last ice retreated: the North Atlantic birds which stem from an Atlantic refuge that moved into continental Europe, and the Aralo-Caspian clade which came out from an area that was also a glacial refuge around the Caspian and Aral seas, where the ancestors of the current Caspian gull evolved. Our herring gulls in Britain came from the Atlantic clade; it seems, though, that some herring gulls also feature, confusingly, in the eastern clade.

The herring gull complex of birds (including lesser black-backed gulls and other – now separate – species like yellow-legged and Caspian) were thought for a time to form an exemplary *ring*, a taxonomical explanation that allowed populations of related birds to have a more intimate, interbreeding, relationship with those living nearby compared to those (though still related) living far away, in geographical terms and in distant evolutionary time. The herring gull ring was thought to circle the Northern Hemisphere, hybridising where they overlapped but meeting – where the ring closed – as two biologically isolated, different species. Herring and lesser black-backed gulls were believed to be this: living at two ends of one ring, the same, as it were, but different.

The trays of the Natural History Museum's bird collection at Tring in Hertfordshire, however, have two hybrid birds labelled *argenteus* x *fuscus* (herring x lesser black-backed)

– the very form the ring concept deemed impossible. And there remain many other questions. Many. Among them, the ramifications of our understanding of yellow-legged and Caspian gulls as full species (Caspian gulls may well be closely related to lesser black-backed gulls) and the polar bloom of several northern white-wingers, like glaucous and Iceland gulls. But while we fiddle over what to call their connection, the gulls are hybridising, their populations are dynamic, their ranges shifting, they are genetically and geographically active within species as well as across bird borders, and they remain in many cases fiendishly hard to tell apart in the field.

★

Pigeons are smutty grey, sparrows are dust balls. The most obvious thing about gulls in built-up settings, along with their size, is how much whiter they are than anything else in urban nature. They also seem able to keep it that way.

Adult gulls are white for a reason. They exhibit the same general patterning of plumage as most birds that hunt from above the sea surface on underwater prey – their white ventral colours countershade against the pale sky and their prey cannot see them. The opposite is the case for other birds hunting above the sea surface. Pursuing a shoal of fish, gulls look like a rotary washing line on a windy day. White birds are very visible above the water, and I've often seen gulls pull in other gulls and gannets and other seabirds.

Gulls are further adapted to sea life. They have more orange and red oil droplets in their eyes than many birds. This helps them see long distances through sea-fret and spindrift. They also have high-functioning salt glands that allow them to live in and off a saline world.

Among herring and other large white-headed gulls there is a 'remarkable uniformity' in behaviour, so says the *Handbook*. The *Dictionary* lists some of the visual and auditory signals shared by all:

> Among the most important, widespread and conspicuous displays are the silent Upright, indicating anxiety or aggression, the aggressive Choking, the begging Head-toss, and the Long Call, a usually ringing and far-carrying vocal utterance, accompanied by complex posture and proclaiming a readiness to interact either aggressively or amicably.

Having much in common across their family, it is not surprising that birds sometimes jump their genes to breed with other species, and that their hybrid offspring grow up fertile. This is vexatious for those keen to separate the birds into species, and suggests that we need a species concept much more than the birds. With the right mate, however, gulls are 'overwhelmingly heterosexual and monogamous' (*Handbook*), although, perhaps owing to a shortage of males, some herring gulls and other species have been seen forming female-female pairs or trios. Where there are enough men, pairing for life is characteristic, but divorces have been reported between failing birds. Most large gulls first breed in their fifth year. By ten, the birds are good at reproduction and stay that way almost to the end of their lives. Herring gulls often reach twenty, and some pass thirty.

Early in the breeding season, the birds gather near their nesting places and seek to re-form existing pair bonds or, if necessary, secure new dates. They then spread out across their colony. There are choking-displays to endorse nest-site choices, there is nest-building, there is courtship-feeding, there are bouts of copulation, there is rape, there are fights. Think of a city, a battlefield, a dance floor, or all of these at once. Almost every activity has its designated noise. At the nest, adults are territorial and will defend their space, a wider area around, and the air above. Attacks on other gulls are commonplace, though severe wounds are rare. They do also manage to sleep – on average for twelve hours a day.

Herring gulls often lay three eggs and, in good times, will rear three chicks. Occasionally they breed solitarily but they are, like almost all gulls, happier when colonial. They will, as I saw in Bristol with Peter Rock, also share colonies with other gull species.

In the 1950s, Esther Cullen did experiments with the eggs and chicks of herring and lesser black-backed gulls, placing them in the nests of cliff-breeding kittiwakes. Lesser black-backed chicks weren't scared of any abyss because they're normally reared in flat places, while herring gull chicks, which usually hatch in cliffy environments, were programmed to avoid drops and edges. But, as the kittiwakes' ledges were narrower than those used by herring gulls, when the planted chicks went on little walkabouts, there were 'disastrous consequences'.

Almost every large gull – in adult summer dress – has a yellow bill. When a gull is breeding, its bill sports the red gonydeal spot. New-hatched gull-life is dependent on this: pecking at its parents' red spot provides the chick with a regurgitated meal, and in this way the dependents call the shots. At about a week old, the fattening chicks begin to walk about. At the same time they learn the sound of their parents.

Wandering young must learn not to move too far. Adoptions are known but cannibalism is commoner.

Many herring gulls in England, at least, are sedentary and don't go far from where they spend their summers. Young birds move furthest, though they often eventually return to breed where they hatched. Adults often stay close to their breeding sites through the year. I read rings on some adult birds on Guernsey in 2017 – none had been seen anywhere else but there. Scottish breeders, like the Scandinavian *argentatus* subspecies (a winter visitor to Britain, especially to the east coast), are more mobile and move south for winter warmth. Lesser black-backs are more prone to move but are in the throes of changing their habits.

★

Gulls are great eaters. They are also genius *goers-after* of food. 'Predator, scavenger, food-pirate,' *Birds of the Western Palearctic* says, 'taking almost anything available of suitable size, texture, etc.' Many studies offer a similar observation:

> In north Wales, stomachs and pellets of [herring gull] chicks contained mostly grass, fish, and invertebrates (marine and terrestrial); regurgitations also contained skin of a ray, conger eel, cow's placenta, sheep's intestine, earthworms, and a tomato sandwich in a plastic bag.

On rubbish dumps they steer by smell: I've watched gulls feel for buried treasure, take it in their beak and, fleeing their piratical brethren, manoeuvre and palpate what they have grasped, testing and tasting the item as they fly. My friend Patrick McGuinness has seen gulls in Caernarfon opening rubbish bags outside hair salons and eating the hair. In arctic Norway in midwinter, I've seen herring gulls capitalising on the turmoil in a fjord after the arrival of groups of humpback and killer whales hunting herring: when the whales freaked the fish, the gulls cleaned up. On Guernsey, I watched a conference of herring gulls alternate between slurping ice cream from discarded tubs on a harbourside (exploiting a newly discovered meal) and flying up above a concrete slipway to drop and crack open some shelled item pulled from a rock (exploiting a newly discovered access to an old meal). Others have seen herring gulls drop frogs and rats from heights.

*

Niko Tinbergen's mid-twentieth-century study of herring gull behaviour, fieldwork conducted mostly in the Netherlands, laid down much of what we understand about the summer life of the birds. Looking at gulls in their breeding colonies 'enthralled him' (Peter Bircham in *A History of Ornithology*) and propelled him to prominence among the founders of ethology. Observe first, he would say, then wonder. Ordinary life was as worth studying as anything exotic: 'house sparrows are as interesting as golden eagles.'

In *The Herring Gull's World* (1953), Tinbergen describes his now famous bill spot experiments – presenting chicks with various coloured and marked shapes and seeing what they pecked at – but also annotates as much of wider herring gull society as he could. Doing so, he describes an elaborate and interwoven sequence of encounters among the gulls that allows for a kind of combative life (the pressure to breed successfully and secure a genetic future) that is able to proceed with little or no bloodshed.

I learned of Tinbergen at the same time as reading William Golding's *Lord of the Flies* (published in 1954), and I remember slipping between the books and their thrilling and terrifying sociological insights: two societies on seashores coming into being, living their own rules, breaking and remaking their lives accordingly; with individuals struggling, as their diverse personalities and selfhoods came under collective coercion, towards what might or might not be a common good. The gulls in the Dutch dunes mostly refrained from killing one another. Golding's boys on the beach didn't get off so lightly.

Paul Celan, the troubled and troubling poet of 'Death Fugue' and other dark works written in the shadow of the Holocaust, knew of Tinbergen's studies and wrote 'Möwenküken' or 'Gullchicks' about the 'pregiven inheritance' that operated on the birds. He used it, darkly, to talk (in Pierre Joris's translation) of humans going down 'both time and eternity's / wrong / pipe.'

★

In 2015, after taxonomic revisions split the circumpolar herring gull into several species, BirdLife International estimated there were between 2,060,000 and 2,430,000 individual herring gulls, *Larus argentatus*, in Britain and Europe. Their numbers had probably peaked in Britain around 1970. The total population doubled between the Second World War and the 1960s. Across the UK, a 48 per cent decline in the species was recorded between the bird survey years of 1969–1970 and 1985–1988, a 13 per cent decline followed between 1985–1988 and 1998–2002, and then a further rapid decline of 33 per cent took place between 2000 and 2011. Despite these losses, the population is still believed to be higher than it was in the early twentieth century.

Most of this decline has affected coastal, more traditional, gulls. Birds that live more directly off people and close to them are surviving better. But overall the herring gull has lost out. We've liked to shoot it for sport, eat its eggs for supper, wear its feathers to adorn ourselves, and more recently we've decided it is a predatory pest. And now a truly modern conundrum has hatched: the species in decline is still regarded as an invasive nuisance.

In the winter of 2017, Gulliver, a herring gull on Jersey, received a death sentence. Its crime was being 'friendly', said John Pinel, the island government's principal ecologist: 'unfortunately that manifests itself in dangerous behaviour'. Hats had been knocked from heads. The council proposed to shoot it.

*

A pair of herring gulls that frequent my local park in Bristol are regular foot-paddlers. On damp winter grass they run on the spot to bring worms to the surface. I once tried to make a little film of the dance on my phone. On the result, you can hear Lucian, my son, out of sight, hurrying me on by pointing out that, engrossed by the quick-step in front of me, I was also filming a busy children's playground with its own paddling pool and swings.

I stopped filming, we walked home and, talking of dump-life and gull-life, Lucian, who was studying anthropology at the time, suggested I should look at the 'rhizomatic' ideas of Gilles Deleuze and Félix Guattari, as formulated in their *A Thousand Plateaus* (published in English in 1987). The world, the Frenchmen say, is organised in our minds like a branching tree, or a deltaic root. The tree is 'the image of the world'.

Knowledge is understood *dendritically*. Systematics works like this, as in the tree of life, but so does political power. We tend to connect everything hierarchically – the trunk and the branch, the deepest root, its furthest fingers – and we are the outgrowth of older things, previous lives and yesterday's thoughts. But nature – natural non-human life – Deleuze and Guattari say, doesn't live like this, not even the trees, that we apostrophise as great rooted blossomers. Nature doesn't depend on such 'pivots' but lives and grows *rhizomatically*.

The evolutionary tree, as the ornithologist Jeremy Greenwood says, is 'not three-dimensional but multi-dimensional'; it needs to show the work of time and connectivity while also seeking to place more primitive species before more advanced ones, and to put closely related species nearby one another. Real trees are webbed into life across a thicket or a forest, not just rooted to a single spot; many organisms are actually two life forms, not one; lichen is a meeting of algae and fungus; we are co-evolved and our bodies are ecosystems – without gut flora we'd be dead.

Perhaps we should think like Deleuze and Guattari and know nature as anti-hierarchical, refusing a single source and growing instead from countless seeds or multiple spores or as rhizomes spread, without a mother-lode or taproot. I'm not sure this is an accurate account of the French brains – their thoughts are (as rhizomatic thinking perhaps should be) hard to follow – but I liked the play that their ideas, glossed by Lucian, set free. A dump might be a rhizomatic mattress where every junked fragment can connect; gull speciation might be construed as more rhizomatically adjacent than hierarchically dendritic. Perhaps I was being over-enthusiastic, but the talk got us home, and I especially enjoyed doing it with some shared genes.

A Peck of Dirt

There are recurring givens among the phantasmagoria of commodities on a landfill. I noticed four at Pitsea, on the Thames shore in Essex. On nearly every outing there, I either stepped on a dildo, tripped on some indeterminate grey cabling, was detained by the pleading button-eyes of a dead soft-toy, or kicked a book by Ranulph Fiennes.

Locally, the dildos are anticipated. Other gullers at Pitsea had spotted the same trend. Some picked them up and waved them at their friends. Tip veterans barely registered them. Their Caucasian pinks were a similar shade to the sickly neon-coral grime that skins the whole dump. It is hard to identify what this is. A generic residue, or distillate, of all that has been tipped? Or some emergent fungal sporing?

Twice at Pitsea I found a book by Sir Ranulph, knight-adventurer, with damp pages and a shrinking cover, the whole thickening to some light brick that might possibly fuel a fire in an emergency. Kicking his face by accident, as I hurried after gulls, I remembered the story of his frostbitten fingers – after a mess-up in Antarctica, where he'd lost a glove in an icy sea channel, his fingertips on one hand turned black; back in Britain and, one day, despairing in the Home Counties, he locked his hand in a vice in his workshop and took off the dead digits with a hacksaw. And then, presumably, threw them away.

★

I got out of my car into a smell. Death, I would say. I could taste it too. Sweet and soft and sickly. And the corners of my eyes and my mouth gummed with that grime all day from that first moment. Gulls passed over my head, eyeing our shared route, me on a gritty track, them in the silted air, and I hurried to keep up. Over my waterproofs I put on my prescribed gull-dump kit: a hi-vis jacket and steel-capped boots. Then I found a crowd wearing the same, fifteen men and one woman. I introduced myself.

This, the North Thames Gull Group, is a team of trained gull-ringers (with some supervised trainees) who go catching at the landfill. They have worked Pitsea for years. For a dozen or more Saturdays, from the late summer to the early spring, under the leadership of Paul Roper, the group set cannon-nets on the site and catch and mark (with metal and colour rings) as many gulls as they can.

I got a lift to the top of the tip on my first trip with Paul. He was optimistic. 'There's a lot of geebs about,' he said, gee-bs being *gbbgs*, great black-backed gulls. We passed a scree of loafing birds, mostly herring gulls. 'That'd be a nice catch, I'd take that.'

For the North Thames Gull Group, Pitsea is a way to get hold of gulls. In this handling (identifying, ageing, measuring, marking) a kind of reading of the birds begins (or contin-ues, as birds are sometimes re-trapped, and birds that have been ringed elsewhere, held by others, are sometimes caught or 'controlled'). An understanding of populations and movements is the goal whilst conservation measures are informed by the knowledge that the ringing brings.

I talked to Paul once the cannons and nets had been set to his satisfaction, and while we waited for a delivery of food waste and for hungry gulls. The launch cable ran back to the group's vehicle at the edge of the trapping site, a flattened hilltop of junk close to the working cliff of the tip, where truck after truck dropped new rubbish throughout the day. Paul stood ready, near the firing-box, but broke off during our talk to chase a fox away, to direct a dust-cart, and to instruct the compactor-macerator driver where to crunch and split the delivery.

'Do you want us to not swear?' Paul asked as I switched on my tape recorder.

He began ringing in 1980, and not long after started cannon-netting. He is a commanding leader but also very funny, with a fabulously inventive blue mouth that spares no one and nothing: foxes on the prowl, drivers of dust-carts on the dump, the weather, gulls, gullers.

'Listen, I've got the gunpowder, so I think I am in charge.'

'We've set the cannon-net. The site foreman has put aside a bit of tip for us. The closer we can get to the tipping face, the better chance we have of taking a catch.'

A line of four cannons – black metal pipes one metre long and three centimetres wide – were angled out from old car tyres and weighed down with soil-filled sacks.

'The food we want when it arrives (we hope) will be spread in front of them. A dropper cable joins the cannons, and that's connected to the main firing cable, which goes back to the car here and the button and my finger, fifty or so metres away.'

With the permission of the tip bosses, Paul keeps an old Land Rover on the tip. These off-road cars are laws unto themselves. They've left the highway forever. As scrapped or worn-out vehicles they have won a stay of execution, but they must spend their spared life driving over a landscape made from other condemned objects.

'The site people look out for food for us; if anything comes in with black sacks or food waste they will bring it up, and drop it right in front of our waiting net. Then the compactor driver will come and give it a really good mashing so the bags open and the gulls can see the food.'

★

In 1979, when James Wentworth Day wrote in his *Book of Essex*, that 'Essex is becoming the dustbin of London', the dumps on the Thames shore were not his beef. He doesn't mention Pitsea or Mucking or any other landfill. His concern was the sprawl of eastward human development, exiting the capital, new towns, trashy buildings and incoming aliens, all of which threatened his beloved old county. Wentworth Day was one of those early-twentieth-century conservative countrymen-writers who saw the end of the world written in the death of the old truths of rural England. Most of these men died before their prophecies could be proved, but Wentworth Day had the misfortune to live on until the 1980s. When he first knew Canvey Island it was a wet marsh for rough wildfowling. By the 1970s, Essex was becoming 'a cheap-jack appanage of suburban London'; a county where everywhere 'the sluttish fingers of the bungalow developer', threatened 'wholesale spoliation'; where once cherished places were being 'brutalised into ugliness'. His chapter on the county-as-dustbin comes before one that rehearses his venerable and county-rooted family tree. The fear of all sorts of swamping, a new muddying of a place and its people, is palpable. And, horrible book though it is, Wentworth Day was half right.

Essex has been a dump for ages. Downwind and downriver of London, it has been shat on for hundreds of years. Bovril boats, so called because of their faecal whiff and excremental colours, used to dump tonnes of London's human waste in the Thames off Essex where the river becomes an estuary and begins its tidal run to the sea. They stopped these drops only in 1998. Other capital crap has often been brought to the shore and tipped there. Essex is good for long goodbyes. In 2011, the Thurrock district (where Pitsea and Rainham dumps are, and Mucking was) had more particulate matter in its air than anywhere else in Britain apart from Nottingham. The dumps along the Thames shore are so many they might merge. Mucking took 660,000 tonnes of London's trash every year. The site could be seen from space. Closer up, the dump workers knew that some of the rubbish they were driving over was sinister enough to melt the tyres of their vehicles.

Bodies continue to be added to the macerated earth of Essex. In 2015, tunnellers digging for a railway came upon the skeletal remains of 3,300 people from Bedlam hospital near Bishopsgate in London. Some were plague victims. The capital had no potter's field, no new room for its old dead, but a place was found for them on Canvey Island.

*

The truth is, however, the dump at Pitsea is dying. Food waste is dangerous to world health. Organic matter entering landfills is believed to add significantly to global methane emissions and could be an important contributor to climate change. Recycling our junked food is better for us. As this becomes more efficient, with much now either composted or burned for useable gases, less arrives at landfills.

For half a century gulls were granted a new way to be. Some landfills could supply a gull with its daily nutrients in less than an hour of foraging. Gulls worked this out within a human lifetime. They have even learned that dumps often don't operate on Sundays and they know they must go elsewhere (they are, though, the *Handbook of the Birds of the World* says, apparently 'confused by public holidays, and wait for hours in vain!').

Large gulls can live for thirty years, smaller ones not much less; they have lived through the boom times and now must work the ruins. PAFs – predictable anthropogenic food subsidies – will soon be no more. On my first trip to Pitsea in October 2015, Paul told me two food-carrying lorries had been put aside for the gulls and us, but 'in the past I'd expect ten on a Saturday morning.' On my last trip in 2016 we had to make do with just one delivery. 'It's no wonder that most of the gulls have pissed off.'

'We're standing on 38 metres depth of waste. Its days are numbered. European regulations insist that within five to ten years there will be no landfill in the UK that will take in 'putrifiable' waste. Brexit won't change that. All food will be composted or incinerated.'

The Clean Air Act that stopped the burning of waste came at a time in the 1950s when by-catch discards of the fishing industry and the practice of gutting fish either at sea or in port areas declined. No longer burned, a new food source appeared in various urban hinterlands; 'the gulls / Wing to the Corporation rubbish ground,' Philip Larkin wrote in a poem from 1962, '[a]ll is not dead.' And so the birds came ashore and lived off our remains. Doing so, they also came differently into our minds. A tidying, of a sort, made the dumps into places for the birds. The cleaning of one part of our house dirtied another, and a *dusting* of food made up the gulls.

The word hurries at you on the dump on any gust of air. *Dust*. You taste it. You cannot help eating it. Leftovers dusting everything; handfuls, should you stoop to the ground, of our exhaust.

'These sites will close, I expect. There'll still be some commercial waste, builders' junk and contaminated unrecyclable material, but eventually everything will be capped over and developed for some other form of land use. Mucking's already gone and Rainham has no putrefying waste going in. Country parks seems to be the plan for these spaces; once everything is sealed underground they can be attractive areas for leisure activities.'

I couldn't hear *country parks* without wincing, but Paul loaded nothing onto the words. He's an ornithologist and said he tried not to be 'emotive'. I wanted him to mourn the tip. He wouldn't.

I don't drop litter. Sometimes I pick up other people's trash, and I have been known to hand back a discarded cigarette packet to whoever tossed it from his car at a traffic light. I've been called a cunt for that. Among the dark-side-of-the-moon revelations of Pitsea, its mad reality, was watching the regular team of gullers there – all committed ornithologists and conservationists – throwing their sandwich wrappers and empty drink bottles to the earth at their feet. It isn't earth, of course; a giant carrier bag stuffed with our filth has split open across 240 hectares. But I took my rubbish home, exporting it along with the grit of the dump's pinkish plasma, its dust's dust, beneath my nails, in my hair, around my mouth and at the corners of my eyes.

'We have theories as to what will happen to the gulls, that's why we're doing this work, our whole study is geared towards trying to see what will happen when the landfills close; in the thirty years we've been working on the tip, we've marked 45,954 birds with metal rings and around 12,000 of those with colour rings, and we hope the numbers of birds marked will help to prove the theories we have. We know the dump has been a big food source for many birds for many years, and it may be that this has permitted larger than natural (or otherwise sustainable) populations, so, we might never equal these numbers without landfill. One of the first concerns is that with the reduction of food, the gulls' winter survival will decrease and the population will decline. The picture is complicated by our discovery that various populations of gulls use the tip in the winter: there are birds from the high arctic, northern Norway and Finland, that come down for a few winter months; there are birds from our own Thames estuary that are here all year; there are birds that breed in the Netherlands, France and Belgium that use the tip in winter.

'Look, we've got a dust-cart coming in now.

'One theory suggests that without landfill, gulls will go towards even more human sites for food, into more populated places and run the risk, if they aren't already, of becoming a nuisance. There is also evidence that they are foraging further afield. We've had increased sightings from French landfills, which aren't closing like ours.'

A hopeful voice: 'I can see some McDonald's wrappers!'

'Right, everyone back behind the vehicle.'

'This sort of waiting game happens when the weather is warm and the gulls aren't so hungry. They're loafing now. We could do with a compactor running over our waste again.'

Waiting for the gulls, I scanned the hi-vis jackets and talked to some other gullers. There was a postie whose hobby was ringing at Tring reservoirs and who was the only woman on the dump that day. There was a PhD student working on urban herring gulls at Billingsgate Market in London, who described his principal study area as a car park. And there was Andrew Tongue, who was researching how flame-retardants, which are in all sorts of man-made items, might accumulate in gull's bodies if they spend much of their feeding lives in the company of wrecked and ruined human consumer products, where the poisonous chemicals persist. Landfills are reservoirs for these. They also lurk in birds' eggs.

Not everything a gull takes from a tip is good for it. Decades of working through our rubbish have made them cruelly useful bio-indicators of various environmental calamities. The feminisation of embryos as a result of DDT exposure was first noted in gulls in the 1980s. And, more recently, gulls' picking up and eating plastic nurdles and other petro-morsels, apparently as aromatic as anchovies, has been a signal of a global plastic crisis.

The gulls at Pitsea weren't coming.

Paul carried on talking.

'The seeming chaos of birds does have an order, I think; we're just not clever enough to see it, but marking them helps us understand. These gulls are living through population dynamics, climate change, continuing evolution; it is all happening, and we're just scratching the surface. Colour-ringing the birds has changed what we know. Using only metal rings achieved five foreign recoveries of lesser black-backed gulls in twenty years; within two weeks of colour-ringing we'd doubled that. With a metal ring you pretty much have to rely on dead birds being found. But all sorts of people notice colour rings.

'It is colour rings that have told us a lot about how the various herring gull populations work. We've got regular resident gulls that use the tip all year round, these are often not-yet-breeding sub-adults that live along the Thames estuary; then there are adult birds in that population that move around a bit more, they might breed in the estuary, but we may get sightings of them down on the Channel coast or on the Essex shore, they wander about mainly in the autumn and spring and are back here in winter. There's another population, what we call the North Sea Triangle, herrings and lesser black-backed gulls, that breed in Holland, Belgium, France and some of the UK east coast colonies (at Landguard and Orford Ness) and they all use the tip mainly in the summer and autumn, when feeding young and then when building up to migrate. Then there are northern herring gulls, the *argentatus* subspecies, that we have found are loop migrants, breeding in north Scandinavia, moving southwards in August and September, getting to the north of the UK in November and December, and to us on the Thames from mid December to January; they don't stay long, some are back in Norway as early as the second week of February.

'There's birds coming over now, they want to feed; that's the feeding call.'

I wrote *ahhhhh, oooooh* in my notebook.

'If the compactor would give us a roll, he'd pull the loafing birds up off the tip. It's a lot about timing. He's not tuned in quite as we are; he's probably playing a game on his phone. But the tip guys are fantastic. They are a rare breed – Shaun Taylor, Dave 'Winky' Winger, Stuart Crane and Peter Budd – and these drivers – Alan and Mark (I've never known their second names) – are great people. They all think we're mad and call the gulls *shite hawks* but they've also been amazed when we tell them where these birds come from and go to.

'Now we've got the fox back too. Being very annoying. Dave! Chase the fox off! We're buggered until we chase him right up the hill. As far as you can! They're so tame. Some of the drivers feed them sandwiches. I threw a bit of wood at one earlier and it ran towards it thinking it was food.

'Nah, we've lost them, they're off back to the main tip.' Paul returned to what we know of herring and lesser black-backed gulls.

'They breed in mixed colonies, they have similar patterns of behaviour; but their migration strategies do differ – some lessers are still wintering in Spain and Portugal and all down West Africa but even that's changing for some of the birds. The species was once a true migrant. In the Channel Islands they still are – they only have two or three in the winter there. Dutch and Norwegian birds still migrate. But the British population has changed and many stay. We don't know why – it isn't to do with food availability on the tip; we don't get many in the winter here. They are kept in the UK by something else. Climate perhaps? Maybe another food source? No one has worked that out yet.

'Here's some looking semi-interested.'

And they were. Gulls flew over the compacted yards of our allotment. The mashed-up waste was working its magic and birds were being pulled down. They landed and began feeding in a hungry saturnalia. I glimpsed a tug-of-war over a naan between two herring gulls.

Paul saw his moment. He pressed the launch button and the net cannoned over the gulls. They lifted as one as soon as it rose above them. A few got away but hundreds were caught. We all started running towards them.

'Get the geebs first!'

Close up, the catch was extraordinary. The surface of the dump appeared to have come to the boil. A teeming grey and white shoal seethed and flailed under the net. It looked like cornered fish or wave caps or a mixture of both. Food was now meaningless; escape the only concern. A sandpapering sound came from below the net and great agitation. Some were pushed by the webbing into the trash, some were squabbling with each other, and some at the edge of the catch were working their way from under the net. We spread out to stop any further escapes and began extracting the birds.

Because we had caught so many – more than 400 – Paul adapted the usual protocols to ensure the birds' safety. The most desired and largest, the geebs, were bagged first, then lesser black-backs, then anything with a ring already on it. One herring gull had been ringed in Norway; another was Dutch. A few Pitsea birds were re-trapped and had their ring numbers noted quickly at the side of the net. Meanwhile sacks were filled with the new gulls. We had to work fast. The dump below the birds – the latest pungent layers of feculent accumulation – was foul with food juice. Panting birds were checked and warm feet were monitored. A carnival was being tidied up with great care.

Holding a gull is something. At the net, experienced ringers untangled one bird at a time from the squirm, usually positioning it upside down and taking both legs in one hand and using their other to grasp its wings and neck. Novices like me approached the scene, careful not to tread on any bird, carrying an open hessian sack, so the extractor could pass the gull directly into the bag. I tied the string at the neck of the sack with a double knot, and carried it and its bellying cargo to the strip of plastic from where the cannon-net had been launched. Three or four black-headed gulls are allowed to share a sack; larger gulls have sole occupancy. On the plastic, some sacks subsided and remained still until they were picked up again for processing. Others vigorously strove for freedom. From within a dark, confined and alien space, some gulls managed to manoeuvre for several, sometimes a hundred, yards. We kept an eye out for any truancy, and a member of the team was allocated to oversee the bagged birds. Tying one bag, I watched as a great black-backed gull escaped from another, courtesy of a worn sack or a bad knot. People were always fetching back wandering sacks. Away from the gathered heap on the plastic strip the sack could be any old junk. And the foxes are bold.

Once the whole catch was extracted the team moved to process the birds. You take a sack, untie the knots and open the neck just wide enough for a hand and an arm. Inside you sweep about, as in a bran-tub lucky-dip, but cautiously for always and not far away, in your mind and for real, is a weapon, inches from your fingers, like a shark's maw or a scorpion's tail: the unseen but absolute centre of the bird loaded into its bone-hard beak.

As with extracting the birds from beneath the net, taking the gull's feet first in the sack is best. They are like parchment, or what I remember of my grandmother's hands, chalky dry and oddly soft but tipped with claws that are not. Next you must reach behind the bird's head with your other hand to grasp its neck and wings to stop them flailing as you pull out your prize. Black-headed gulls are small in dump-gull terms but still have a powerful draw to their wings and a bloodthirsty penknife out front. As I got bolder, which meant holding on more firmly, I could just about manage them into the grip of one hand.

Birds' feathers always feel more resistant to the touch than you expect: they are tough like a raincoat over softer woollens, a shifting outer skin. A dust comes from them too; feathers breathe and crackle. It pays to hold on tight. Large gulls were often too brawny for my neck-grasp and, with them, having got some feet to grip, I used the sides of the sack as a corset for the bird's wings and drew it out slowly, reaching with my other hand, as the bird slipped from the hessian, to secure its head and to keep its beak away from my face. Ringers handle large gulls somewhat like people handling snakes. Sometimes, my own reach was too slow or my grasp too cautious and, as the bird came from the sack, it tried to fly. Once, losing my neck grip, I was left holding a great black-backed gull upside down by its feet while the huge vans of its wings milled and thrashed and it coiled its neck and head and beak upwards to bite me. To stop the chaos, I had to draw the monster close and cuddle it.

Whether emptying the sack was a smooth operation or not, there followed an action that I learned from watching other ringers and which I think unique to gull ringing: continuing to hold your bird by its feet in one of your hands, you release your other hand's grip on the gull's neck and swiftly bring its body towards you and wedge it under your arm, its feet and head facing up, its back to the ground. It should look as if you have furled a living umbrella and kept it close to you by clamping your arm to your body. The gull's feet are now accessible to your left hand (if you are right-handed) and the rest of its body tucked safely away in your armpit save for its head and beak, which protrude from your back just below your left shoulder.

Somewhere during this pantomime you should have identified and aged the species you are handling. I always double-checked my opinions with others: a bird close-to is quite unlike a bird down a telescope, and the elaborate and protracted moult sequences of large gulls turns even a mass trapping of herring gulls into an extraordinarily diverse collection of individuals. But once identification and age were confirmed, I could ring the bird, joining a queue to collect a colour ring and a metal ring and to borrow a pair of pliers. Black-headed gulls take metal rings that are reasonably easy to close about the birds' legs; the larger gulls require a stronger grip and greater dexterity. Colour rings are glued shut rather than squeezed. On one cold winter session, a little gas stove had to be lit to warm the plastic rings enough to make them sufficiently pliable.

Birds express their personalities as they are handled. Some get irate, others stay calm, but even an irked bird can be contained by skilled hands. Confident processing by an expert seems to secure the bird's assent. Flying remains their intention, but they agree to the temporary halt and inspection. It is beautiful to observe a good ringer handling a gull, moving the bird without fluster, able to extend its wings by the full stretch of their arm to assess moult and age, to spread its tail looking for signs of deprivation or stress recorded in feather fault-bars, to look closely at its iris, in the oily globe of its bright eye, and then to seal the rendezvous by the deft application of the rings. The weight added to a bird by any two rings is calculated to be the equivalent, in human terms, of something between a wedding ring and a wristwatch. Perhaps it could be a more literal equivalent too: the badge of an encounter and a timing device that might be read down the years.

The birds aren't harmed, but almost every ringer on the tip had bloodied hands from beak-bites. As we queued for rings, you had to watch out for the gull protruding from the armpit of the ringer in front of you. Sometimes a conga of birds and people were linked together as one great beak after another jousted with a neighbouring gull, or snapped and drew blood from the hand of the next ringer in the line. Every time I left the dump my hands were scored with beak marks and broken skin. The imprint on my thumb of the bill of one great black-backed gull stayed for a week.

On one of my days at Pitsea, I was designated scribe and sat on an upturned plastic box, my back to the cannon-carrying trailer, with a folder of record forms on my lap. A queue of odd couples, ringers and ringed, lined up in front of me. The ringer read out the unique identifying digits and letters of the rings. The ringed had little to say; the gulls generally remain silent throughout their meetings with men. I wrote down the details in the register: the species, the ringer's initial, the ring numbers, and the bird's age which is also given as a number.

'83 is a 3.'

'97 is an 8.'

'22 is an 8.'

'23 is an 8.'

'10 is a 7, sling it.'

'99's a 3.'

'24 is a 3.'

When the data's captured, the birds can go. They must be thrown back into the air. Foxes will take any gull, especially black-headeds, which wander off on foot to adjust their plumage after being handled. Most shit on departure, and dangle their legs, figuring their rings, for a second or two, before raising their undercarriage and lifting themselves back into their element.

I watched them fly, the gulls I'd ringed: the first time in human hands for them, the first grip of living gull feathers for me, our meeting underwritten by, and only possible in, the squalor of Pitsea. It was getting dark when we finished. I walked off the dump feeling high on eight hours of gulls and trash. Hundreds of hovering, then handled, birds with rubbish at their feet. As I came close to the edge of the site, I saw beyond the landfill for the first time all day. From a last ridge before the entrance yards, you can see the Thames estuary, sunken and twisted with countless creeks slipping towards the sea. Pylons marched everywhere. There was another dump-cum-riverside-park along the shore and an oil refinery nearby. Mist was forming over the shallows and the grey-green sea grasses. A few moored pleasure boats tilted on the low-tide mud but everything else seemed intent on moving.

The day had been grey, with a deadly sky, but it started and ended shining bright. Dawn was lit by the thrown light of the last of sleeping London and the metallic pins of planes falling slowly west towards Heathrow. It was the same at dusk, except the sky was busy then with gulls – our ringed birds among them – heading out east towards the widening estuary to sleep. Like running metal stitches on a gold cloth, they marked it as they went.

*

The good catch didn't change things. In the autumn of 2017 the North Thames Gull Group couldn't work at Pitsea. The food waste coming onto the landfill was too meagre to make trapping worthwhile. The Group – still wanting to fly nets over birds – began investigating goose-ringing options elsewhere in Essex.

Lump

As soon as our own species stepped away from the teeming continuum of the surrounding world, we started naming its variousness. We've continued, and being endlessly wrong about it hasn't stopped this naming of nature from making some of the great songs we have sung. In the Biblical account of creation, prior to the fallout over the apple, the separation of humans from other animals needed to be asserted. The required fencing was to be achieved by capturing all other life forms with names. Adam in the Garden, even before things went haywire with knowledge, was given an Eden project of his own: the job of naming every beast of the field, every fowl of the air: 'whatsoever Adam called every living creature, that was the name thereof'. 'Naming the Animals', a poem by the late Anthony Hecht, develops the story. Adam, new – understandably – to the game, is bewildered by his task. Shyly addressing a cow, he ventures to call it 'Fred'.

Fred the cow and Adam's efforts underline the paradoxical effect of naming nature. Organising the world systematically with man-made words, representations or symbols brings wildlife closer to us. Other forms of life exist and live, but we notice what's going on and do things to it. That might be one definition of *Homo sapiens*. Differentiating every hummingbird in the Americas by a scientific and a common name raises each hectic blur of metallic colour into a humanly graspable object: the birds exist and now we know them to exist. To be aware that there are 340 varieties of hummingbird, where once there were thought to be just twenty species, enlarges both biodiversity and our minds. But, as we discover quite how profuse life is, we also learn how each bundle of feathers cares little for our taxonomical attentions. Naming gives us some purchase, but it also teaches how nothing in nature is for sale, how nothing outside us – animal, vegetable or mineral – can be truly ours.

Contemplating the findings and decisions of taxonomists and other organisers of the wild is a good way to keep fresh the fact that evolution is far from finished. Something about its description by Darwin and other bearded Victorians has meant that evolution is often linked in our minds to the time of its discovery. That is wrong. Nothing is fixed in the world of the gulls of Britain or of any life anywhere on the planet, including ourselves. We are living our own evolution. It is as *relentless* as ever.

Every taxon has its own signature and, as we discover more and more species, we are having to learn to read over and over again; our naming, therefore, can only be a holding measure, a place to put things and a means of finding them again, the truth for us for now. Taxonomy has its own taxonomy and among its terms are waste-basket taxa – for organisms that do not fit elsewhere – and Lazarus taxa – for those that disappear and then come back. If an animal or plant is described as new to science, it means we have pulled a species into our ken from the riot of life, but it doesn't necessarily mean the riot itself has got rowdier. New to science is not the same as new to nature. The known world is not a match for the world.

Nature's writing, therefore, is not the same as nature writing. You might go to your window in the morning, open the curtains and think how neither one blackbird you see knows that it is a blackbird, nor one tree cares that it is an oak, an ash or a lime. Not one. And yet the blackbird lives as a blackbird not as a blackcap, the ash is an ash and not an alder. They don't know their names, but they live who they are. And we are right to tell the difference because the difference tells. And this truth is imaginatively stimulating as well as intellectually invigorating. How much better to know that instead of generic *little brown jobs* flying away from us, there are rock pipits and water pipits and tree pipits and meadow pipits. How great a day it was when Gilbert White's repeated walks around his parish at Selborne in Hampshire in the 1770s prompted him to realise that there wasn't just one but three kinds of leaf-green warbler among the spring foliage. Three species, not one: chiffchaffs, willow warblers and wood warblers.

Careful looking lies behind this enlarging of life: the only way we know there is a rock and a water pipit is by human observers' close attention to the birds' reality. Our species has been doing this for a long time: one of the wonders of cave paintings and rock art is the accuracy of the depiction of animals. We might not know whether the lions or the wild horses represent a shopping list, a field note, or a frieze of totemic beasts to inspire shamanic dreaming, but we can immediately recognise specifically what the creatures are. They are named in ochre.

We might call this nature-capture a species of love. Not everyone would agree. Post-humanists debate our appropriation of nature. Some think naming is a quasi-colonial action, a possessive anthropo-something grab after other species; some think that attempts to organise the order of life and to understand it to lose sight of its deeper truth and dilute its organic magic. Allergies to taxonomy are as old as any organising intelligence. In the early nineteenth century, John Clare's poems and prose writings describe a remarkable sixty-five first bird records for his home county of Northamptonshire. Their names brightly mark all of his writing but he was hostile to the idea of scientific study: 'I love,' he wrote, 'to see the nightingale in its hazel retreat and the cuckoo hiding in its solitude of oaken foliage and not to examine their carcasses in glass cases.' Another poet, John Keats, was furious with those who, as he put it in 'Lamia', would 'unweave a rainbow'. The men at Pitsea would disagree, I think. If you could unweave a rainbow mightn't it appear even more magnificent? If you can dial into a Norwegian great black-backed gull and know where it hatched and when it first crossed the North Sea and how many times it had fed on Essex junk, wouldn't that make the world more interesting? More lovable, I would say, too.

★

As the British bird list got close to six hundred confirmed species, and then arrived at this (long anticipated) point, and passed it, I spent some time talking to the poet Paul Farley, a friend and colleague, as he scoped a possible project: could he write a poem for every bird on the list? Paul is a birdwatcher. He's catching up with me. As I was stuck indoors one day, he texted photos of the waxwings he'd seen near his home in Lancaster. We interrupted a BBC interview we were recording once so he might twitch a glossy ibis in the company of three great white egrets at Wicken Fen. We watched it crossing a milky sky, and Paul nailed the silhouette of the ibis's dark and slender flight. It looked, he said, like a heroin user's tarnished tea-spoon. As he got close to finishing a poetry sequence that updated Michael Drayton's seventeenth-century county-by-county gazetteer, *Poly-Olbion*, an epic poetic mapping of the matter of England, Paul had toyed with his idea: six hundred poems for six hundred species. He'd banked some already – memorable and successful lines on house sparrows and a grey heron.

I could imagine a poem about glossy ibises. There was plenty of stuff going on around those birds and they were distinctly their own thing. I watched a flock of twelve working a drying marsh on the edge of the Severn estuary in Gloucestershire in the spring of 2007. It looked like it had rained umbrellas. But six hundred poems? Six hundred birds? A new parliament of fowls? How then to make a Radde's warbler poem identifiably bright and different enough from a dusky warbler poem? The yellow-browed warbler's moment separable from a Hume's leaf? A Siberian chiffchaff from an Iberian chiffchaff? Most of the six hundred birds are stragglers to Britain; only two hundred or so species breed or habitually visit; the bulk of the list is made up of accidentals. Often these are also hard to tell apart in the field and many of them have barely gone in to anything other than the identification and classification departments of birders' brains in Britain. They don't occur anywhere else.

It seems sure that we should expect a long wait for the first British poem about an Azorean gull. The six-hundredth species admitted to the national list struggled to make much noise. It was a yelkouan shearwater seen off the Devon coast in July 2008, and allowed as a British bird in 2016. It lives normally in the eastern Mediterranean and the Black Sea. The Balearic shearwater lives further west in the Mediterranean and is very similar. For a long time both these species were regarded as subspecies of the Manx shearwater.

Nonetheless, Paul is writing and some gulls have darkly got to him. He sent me this:

Great Black-Backed Gull

The tide keeps bringing everything you need.
The tip is like a slowed down sea to gulls
who trawl behind the trucks or dip in strong
kinking glides above the ribbons and shreds
of dross, the spume and swell.
                    Taken to see *Jaws*
at the ABC, the robot Great white shark
made us jump, but later came the slower thought
of real Great whites cruising the seas of the world
while I lay in bed. *What we are dealing with here
is a perfect engine, an eating machine...*
Now, a landfill lubber with binoculars,
I pick out Great blacks from the smaller gulls
above the waste where, fathoms deep, the shark
still swims among the wreckage of who we were
forty summers ago. They're such powerful birds
and the tide keeps bringing everything they need.

The Birds

Daphne du Maurier's short story 'The Birds' is mostly known as the inspiration behind Alfred Hitchcock's film of 1963. Although the gullage in the story has got obscured by its filmic copy, it is actually much more interesting.

Forced to show birds on screen as malign attackers, Hitchcock could only do stagey work. The crudity of the available footage made for clownish near-comical scenes scored with FX squawks.

We have a long wait even for that. One hour into the film the birds are still a subplot. There are pet lovebirds, bird puns and flirty talk; there are sickly chickens; finally there are some massing corvids, a crow parliament on a climbing frame, and then a gull ambush at a children's party. But the attackers could be from outer space. The turning of the birds, familiar and local species, into murderous menaces is hardly questioned. The film is dumb about its title subject. 'Don't they ever stop migrating?' someone asks. Perhaps we are to think the outsider, Melanie Daniels, has brought with her into the small town a sex-infection or strain of bird flu: the gulls first attack as Mitch Brenner starts to weaken and fall for Melanie's charms. Pretty soon it is all screech and flail with the birds mugging their own roles.

The film does have a good guller and she gets her moment. Mrs Bundy, an elderly lady in a beret taking shelter from a downtown gull attack, is having none of this scenario. The birds' brains aren't big enough to cause such mayhem, she says. It is mankind, rather, who insists on making life difficult on this planet. Competition for dwindling fish stocks will have driven the gulls to attack, she says. But nobody cares for experts as the great gull-apocalypse begins.

I tried to identify the species. They are mostly dark-backed, perhaps western gulls. Did Ray Berwick, the credited 'trainer of birds', cast from the east coast or the west? The Disney studios in California worked on this sequence, but the gulls could have been drafted from elsewhere.

The kamikaze birds hang like angry angels above the little port town and its burning gas station, where firemen work their hoses in vain. Melanie takes shelter in a phone box as gulls descend and smash its glass walls. There is a well-done horror medley: claustrophobia, the fear of being buried alive, and the agitation of having a bird entangled in your hair. But, overall, it is hard to watch this scene without being distracted by the strenuous effort that has gone into making it.

Another, much quieter, moment is stronger. It follows an attack by the joined forces of the birds on the Brenner house, when windows are smashed and beaks axe through doors, whilst the pet lovebirds cower and their human avatars play out their roles, the manly man gull-fighting and his solicitous girlfriend comforting her putative sister and mother-in-law in their distress. When the noise has died down, Mitch and Melanie leave, stepping outside into a weird twilight. All is still, but great numbers of gulls and crows crowd around as far as you can see. It is night on a battlefield. There's a lull in the storm. Most of the birds are on the ground. Nothing flies. All are silent.

★

Du Maurier's word-suggestions fare well in comparison with Hitchcock's visualisations. Her story has plenty of creaks, but the bird-police would find much less to prosecute. First of all, this is because it is clear that du Maurier actually watched gulls. Hers, at least in part, is a translation out of nature.

Like several fictional and many real bird-men of the twentieth century, du Maurier's lead character is a wounded and lonely creature. Nat was injured in the Second World War and walks the coast (somewhere west of Plymouth, if I understand the story's compass) alone but observant. A farmworker, he knows his birds and how autumn skies can swirl with them. He detects increasing numbers of gulls. These might be seasonal aggregations and movements, but a tractor driver reports a new boldness in the birds – 'one or two gulls came so close to my head this afternoon I thought they'd knock my cap off!' – and, when the wind howls out of the east that night, birds crowd and scratch at Nat's window. They might have wanted shelter but when Nat goes to investigate they draw blood – 'they flew straight into his face, attacking him' – and, worse, they've also found their way into his children's bedroom:

> He seized a blanket from the nearest bed, and using it as a weapon flung it to right and left about him in the air. He felt the thud of bodies, heard the fluttering of wings, but they were not yet defeated, for again and again they returned to the assault, jabbing his hands, his head, the little stabbing beaks sharp as a pointed

fork. The blanket became a weapon of defence; he wound it about his head, and then in greater darkness beat at the birds with his bare hands.

These first attackers aren't gulls but small birds, including 'blue tits, larks, and bramblings'. Bramblings are among the most innocuous of finches. Did du Maurier really see some latent aggression in them, or were they recruited because they are birds that come to Britain from the east?

This is a Cold War text and was first published in 1952. There is an unseasonal freeze and a harrowing wind. And the birds form a horde or a leaderless mob. A 'madness' has seized them. They have crossed the borders of their own behavioural norms, 'their own flock and their own territory', and leaving their rightful places, they have become transgressive.

'Is it Russia?' asks Mrs Trigg in the farmhouse kitchen.

Nat hunkers into himself and keeps his eye on the sky. It has grown so cold that he cannot bury the dead birds that have crashed into his house. He takes the corpses to the beach but they blow away from him. Then the gulls come. It is as if the drifting feathery bodies have reformed and revitalised:

What he had thought at first to be the white caps of the waves were gulls. Hundreds, thousands, tens of thousands … They rose and fell in the trough of the seas, heads to the wind, like a mighty fleet at anchor, waiting on the tide. To eastward, and to the west, the gulls were there. They stretched as far as his eye could reach, in close formation, line upon line. Had the sea been still they would have covered the bay like a white cloud, head to head, body packed to body.

I've felt that menace, too, on those days of high wind when gulls sit stubbornly on the sea, stoically facing into the cannonade, seeing it out, looking it down; and at those pre-roost assemblies too, where the chaos of the birds' arrival from all quarters is, as it were, magnetised, and all settle on water or mud or field and stare the same way under no instruction.

It gets worse in 'The Birds' when the gulls come ashore. The radio news begins to report that an Arctic airstream is driving the birds south. In London, the sky is blackened with great clouds of starlings and pigeons 'and that frequenter of the London river, the black-headed gull'. The suggestion is that they are all birds from elsewhere, and are therefore malign.

Nat's wife says the government should 'call the army out and shoot the birds . . . something should be done.'

When Nat goes out again it is into twilight: 'It was the gulls that made the darkening of the sky.' A warrior armada spreads in huge formation. A great mushroom of black cloud becomes birds falling to all corners. A knot of crows is bound inland. 'They've been given the towns . . . The gulls will serve for us.' Appropriate malevolence is being apportioned. Nat calls the telephone exchange but no one seems to care. The gulls are keener listeners and mind-readers; overhead, now, they 'waited upon some signal':

> mostly herring gull, but the black-backed gull amongst them. Usually they kept apart. Now they were united. Some bond had brought them together. It was the black-backed gull that attacked the smaller birds, and even newborn lambs ... the black-backed gulls were leading. The farm, then, was their target.

A neuro-taxonomy of the telepathic attackers takes shape: 'the herring gulls … had no brains. The black-backs were different, they knew what they were doing.' The war begins: wave after wave of strikes, each bolder and more suicidal than the last. The birds become their own weapons with 'no thought for themselves'. As Nat reaches his cottage door a gannet dives at him.

The family board up the house, hide in the kitchen, and switch on the wireless. A plane overhead sounds promising, but it stutters, birdstruck, then crashes. Mastery of the air is now reserved for other fliers.

'There's always gas. Maybe they'll try spraying with gas.'

There's a lull. Nat tries to go out and to patch his broken windows with 'bleeding bodies'. The next attackers come down the chimney.

In a second calm, Nat takes his family to the farm, but the people there hadn't read the signs nor had they listened to his warnings:

> Jim's body lay in the yard … what was left of it. When the birds had finished, the cows had trampled him … He could see [Mrs Trigg's] legs, protruding from the open bedroom door. Beside her were the bodies of the black-backed gulls, and an umbrella, broken.

Nat hurries his family back to their cottage. He thinks he sees a navy flotilla offshore but it is gulls massing at sea.

'Won't America do something?' asks his wife. 'They've always been our allies, haven't they?'

Sitting next to the silent wireless, Nat smokes a last cigarette. Small birds peck at the window and hawks attack the door.

Nat listened to the tearing sound of splintering wood, and wondered how many million years of memory were stored in those little brains, behind the stabbing beaks, the piercing eyes, now giving them this instinct to destroy mankind with all the deft precision of machines.

★

Sometimes in a British winter, the north comes knocking and the large gulls regain something of their oceanic heft. Their icy plumage helps: the sharp shave of their heads, their cold eyes watered by freezing winds. When this happens, even regular herring gulls can look as if they deal in more primal truths than what much of Britain offers ('*Sous les pavés, la plage!*' went a slogan of 1968 in Paris, and sometimes the gulls say it too). They look carnivorous. They look big. And the northern species – glaucous and Iceland gulls – that often arrive from the Arctic on the same bad weather further assert the old verities. I would have liked to see the *danse macabre* on 6 March 2017 when, among the polar birds in Britain, two young glaucous gulls were reported to be skimming the length of the west coast and putting the frighteners on everything. One was on Tiree 'feasting on a dead whale'; another at Braunton Burrows in Devon was 'feeding on a dead conger eel.'

The first glaucous gull I saw was notionally more domesticated. It had a name – George. From the 1960s to the 1980s, he spent so many winters between Cley and Salthouse on the north Norfolk coast that people looked out for him and saluted as he passed. I saw him on three or four twitches to see other birds. On these weekend trips, I often slept a few cold hours on a wooden bench in the beach shelter at

Cley (it was washed away in 2013 when the North Sea surged over the shingle). I remember waking up one time and, from my thin sleeping bag, seeing George lumber heavily over the field inland of the beach. A bed tick for me, but I was shivering and the dirty cold white of the bird, like the stained ice at the back of a fridge, chilled me further. There was no black at all on George, but his nicotine tone and urine spots gave him the air of a rough sleeper. 'White-wingers' the birders call the northern gulls, but they make me think of an old Soviet factory ship rusting out at sea and home to various Yuris and Leonids.

It was said that the year after George gave up the ghost, a new first winter glaucous gull appeared at Cley. Boy George they called him.

London Labour and London Poor

It is worth thinking about the absence of gulls in London when Henry Mayhew was writing his multi-volume survey *London Labour and the London Poor*. His fieldwork for this book, using shorthand to record thousands of interviews with street-working people, was carried out in the 1840s, and the assembled whole was published in 1861 and 1862. Throughout the nineteenth century, seagulls barely came up the Thames and, as far as I can tell, they feature nowhere in Mayhew's books.

Gull's eggs were collected for food in other parts of Britain, and were sold in some markets in nineteenth-century London. Over-exploitation, especially of black-headed gulls, in earlier times meant that the trade declined through the century when Mayhew was at work. Otherwise, the birds were beyond the street-level commodification that Mayhew's research centred upon – the world of 'getting and spending', which Wordsworth said was 'too much with us' in his and Mayhew's time. By the mid-1800s, many other birds had landed in London as items of monetary worth. Mayhew records dozens of species, including bullfinches, yellowhammers, Java sparrows, parrots, turkeys, chickens, magpies, nightingales, woodcock, larks and grouse. There were sack-loads of others. Nests and eggs and feathers were also for sale.

Such supplies had to meet multiple demands. Each bird had its price. Some were eaten, others would sing, some would dress a hat, others amuse from a cage. Many were *duffed* – passed off – their plumage prettified, legs and beaks varnished, females painted as males, pale specimens brightened up, local species dressed as exotics.

★

It would be difficult to have read everything of Mayhew. It would, you suspect, take as much time to read him as it took him to make his books. His double-columned, small-print encyclopaedia seems like a labyrinthine business directory. It feels as large as its subject, as big as London, longer than the longest of novels of Charles Dickens, who, being so close in some ways to Mayhew, might have dreamed him up or have been dreamt by him. (Indeed, I might easily have focussed here instead on *Our Mutual Friend*, the greatest novel about London's rubbish ever written, published just three years after Mayhew's study.)

As in Dickens's London, work is the cruel answer to everything in Mayhew's city. It is the only way to live. Mayhew's informants work even as they speak of their work. The book is doubly sweated into life. Its talk is work in itself, patter or voluble effort raised above a crowded street. It is exhausting to hear, and it is often exhausted as speech. I know of no other book that evokes tiredness in quite the same way. What its people report is tiring. How they tell it is tiring too. The prose enacts its own weariness. It also makes its own refuse. Everything stews, strained and dirty. And then, big as a doorstop, the book trips you up. And panhandles. All the *Big Issue* vendors are out at once. All the encyclopaedia sellers.

Mayhew's study is a work of epic taxonomical ethnography. He was, said the historian Eileen Yeo, a 'relentless classifier'. Not only did he take down the words of the working poor of what was then the biggest city in the world, he also set about counting and systematising everyone. His methods were borrowed from biology. His years were Darwin's years, and he had his own interest in the rhizomes of life. He wrote in *London Labour and the London Poor* that one of the most important ways to display data was through 'a correct grouping of objects into genera, and species, orders and varieties'.

Mayhew's organising intelligence pressed on him: tribes were discerned and people labelled; everyone was identified by his or her task. The chaos of the street is sorted into categories of labour and labourers. That sounds utilitarian, but the specificity of individuals was crucial. One person after another illuminates Mayhew's book. His recording was generous. He was interested in types, but he fell again and again for the individual. It is 'Watercress Girl' we hear, not 'A Watercress Girl'. She has never seen – has never even heard of – the green parks of central London, and doubts she'd be admitted were she able to take time off her work and go and look.

*

Mayhew had a problem with people who did nothing, who owned up to no occupation, whose activities seemed superfluous. Confronted by people who looked like this – the categorically in-between – he found a job for them to do, or rather he converted their doing-nothing, or doing-very-little, into a quasi-plausible catalogue entry. It is these people, who turn up everywhere in Mayhew's London, who are *duffing* the work of the gulls in the city. There was nothing to lure the birds upriver because thousands of Londoners had already got their hands on the capital's rubbish.

Trash has a deep and determining place in Mayhew's cosmology. Waste management, in its widest sense, is vital to the story. This begins with the lowest class (Mayhew calls them *low* but was clearly sympathetic to such people). They endeavoured to eke out scraps for a penny or two from what others had decided was useless. Contemplating such lives and such labour makes Mayhew ask big questions. When do objects – or people – cease to have value?

There are dustmen in Mayhew – men in the vanguard of professional waste collection. But they were far outnumbered by informal rubbish collectors. On these people Mayhew performs a kind of rescue anthropology. He describes them as if they were members of a ramshackle federation:

Bone-grubbers and rag-gatherers
Pure-finders
Cigar-end finders
Old wood gatherers
Dredgers, or river finders
Sewer-hunters
Mudlarks
Dustmen, nightmen, sweeps and scavengers

'Pure' is dog shit. Its name alone indicates our classificatory anxiety about its status. It was sold to tanneries, where it was used to cleanse and purify leather. In London, 200 to 300 men were 'engaged solely in this business'. A covered basket and a glove were required, though many dispensed with the glove, 'as they say it is much easier to wash their hands than to keep the glove fit for use'. There were even those who worked fakes and passed 'mortar' off as pure.

The hyphens in the job titles are important. By linking a noun with an action, a haphazard or accidental moment becomes an activity, a gesture is raised into a job, and a desperate grubber after dog-shit is almost unionised. The historian Carolyn Steedman, who has written brilliantly on Mayhew, noticed this: 'reading these passages . . . is like being present at the birth of speech, as if in a kind of space where the gesture creates language, some Rousseau-esque realm of immediacy and plenitude, where if a woman sifts she is a "sifter", and if a man shovels the dust to fill her sieve he is a "filler-in".'

Even with their titles, Mayhew knew that these lowly men – they were almost always men – weren't really doing jobs. They were begging of the earth.

Among the finders there is perhaps the greatest poverty existing, they being the very lowest class of all the street-people. Many of the very old live on the hard dirty crusts they pick up out of the roads in the course of their rounds, washing them and steeping them in water before they eat them. Probably that vacuity of mind which is a distinguishing feature of the class is the mere atony or emaciation of the mental faculties proceeding from – though often producing in the want of energy that it necessarily begets – the extreme wretchedness of the class. But even their liberty and a crust – as it frequently literally is – appears preferable to these people to the restrictions of the workhouse.

Mayhew listens as well to the stories of chimney sweeps, sewer workers and night-soil porters. Each task gets darker, dirtier, fouler. Human excrement was collected at night from communal latrines and cesspools and from the open street. The job was hidden in the dark hours – a night-shift – partly because the work was so nauseous. A *holeman* climbed into a cesspool with a tub. A *tubman* raised the tub when filled. Two tubmen carried the tub to a cart. Mayhew went out with a night-soil team once. He was sick.

★

Dust is everywhere in Mayhew's city. It is another problem for him. He knows that there is no such thing as dirt: it exists – just as Mary Douglas spelled out a hundred years later – only in the eye of a beholder. 'No single item', she said, 'is dirty apart from a particular system of classification in which it does not fit.' But, for Mayhew, dirt, as dust, is the one thing he most wants to define.

Yet he can never fix it. How do you count dust? How do you hold it? What is it? The powdered world? The fundamental raw material? Sediment or suspension? A cast of everything that has lived? That which we tread on – or breathe? That which we are? Hamlet's quintessence?

'Dust hoi!' The dustmen call out, but they never get it all.

Mayhew says that someone once called dust 'mud in high spirits.' Might that be the one? He tries, over and over, to list types of dust. He sounds confident at first. But dust will have none of it. It is antipathetic to lists and to confidence itself. It will not be piled up. It is *vague*. It settles and stirs. We can watch Mayhew's mind in action and see it falter. We know before he's got halfway, that he knows this effort won't hold, that he will have to try again, that matter and antimatter are an uncountable *one*.

A dust-heap, therefore, may be briefly said to be composed of the following things, which are severally applied to the following uses:

1 'Soil,' or fine dust, sold to brickmakers for making bricks, and to farmers for manure, especially for clover.

2 'Brieze,' or cinders, sold to brickmakers for burning bricks.

3 Rags, bones and old metal, sold to marine-store dealers.

4 Old tin and iron vessels, sold for 'clamps' to trunks etc, and for making copperas.

5 Old bricks and oyster shells, sold to builders, for sinking foundations, and forming roads.

6 Old boots and shoes, sold to Prussian-blue manufacturers.

7 Money and jewellery, kept or sold to Jews.

Mayhew then charges at another list hoping to catch its contents unawares: *Street-dirt* must be divided into (a) dirt (b) horse dung and cattle manure (c) mud (d) surface water. Yet even that seems too *thin*. He tries again to parse *detritus*:

1st In a perfectly dry state, so that the particles no longer exist either in a state of cohesion or aggregation, but are minutely divided and distinct, it is known by the name of 'dust.'

2nd When in combination with a small quantity of water, so that it assumes the consistency of a pap, the particles being neither free to move nor yet able to resist pressure, the detritus is known by the name of 'mac mud' or simply 'mud,' according as it proceeds from a macadamized or stone paved road.

3rd When in combination with a greater quantity of water, so that it is rendered almost liquid, it is known as 'slop dirt.'

4th When in combination with a greater quantity of water, so that it is capable of running off into the sewers, it is known by the name of 'street surface water.'

But nothing is coming of nothing. He must speak again. And again.

What is missing in Mayhew's street survey is the mob, the street in riot or what dust does: when individuals are united not by what they do (their jobs), but by what they intend (their minds, their will). Gulls working the tip are like this. They are *public* birds, as political as any I've ever seen. They seem a far cry then from Mayhew's organised London and would be more at home with Baudelaire, whose city streets were contagious territory where everyone risked a jostle or buffet from everyone else. But, who knows, had Mayhew seen gulls in the capital, perhaps he might have found a correlation between *gull* and *dust* and all the grey-brown amorphous anarchies.

Old Birds, Old Books

Mostly gulls are white, in our mind's eye at least, and, often, despite what we know today, they remain associated with the sea. When the BBC took herring gulls off the signature tune to *Desert Island Discs* there was an outcry. People complained that they couldn't see the desert island without the sonic signifier, even if the signifier was incorrect. An engineer put them back.

Gulls were the sea's creatures for a long time. Far out, they didn't find their way into human places, or feature much when early naturalists first began to write about birds they had seen. Gulls were unfamiliar and were figured as wild and remote, keeping cold company with oceans, storms and ice.

In 1966, when the ornithologist James Fisher reconstructed parts of *The Seafarer* poem, written about 1000 AD, as a 'field man's record of a day on the Bass Rock', he translated *maew singende fore medodrince* as 'kittiwake singing instead of mead'. The bird's mournful cry across a dark swell was balm for the seagoing poet.

Gulls long remained tokens of the far-from-home and the storm-tossed. J. H. Gurney was an early twentieth-century ornithological historian, as well as the first biographer of the gannet. In *The Early Annals of Ornithology*, he directs gullers looking for precocious British larophiles to William Turner's

1544 Latin treatise on British birds. Turner commences his account of gulls by reheating a debate as to whether poets or philosophers should be best trusted as accurate observers. He tussles over identifications (gulls, coots, cormorants) with Virgil.

Three lines from Book One of Virgil's *Georgics* describe the status of larger gulls in almost all western literature before the twentieth-century, they are birds of the sea, associated with rough conditions, and seen ashore only after wild weather. This is C. Day Lewis's translation:

> At such a time are the waves in no temper to bear your
> curved ship –
> A time when gulls are blown back off the deepsea flying
> Swift and screeching inland...

Some of the first taxonomies in western culture are found in the Bible. In the book of Leviticus, all gulls are declared unclean. A 'sea gull' appears in the list of forbidden comestibles in the New Revised Standard Translation of the Old Testament. But in the King James version, the same bird was called a 'cuckow'. Does this slip indicate an unknown word or an unknown bird? Is it a mistranslation or a misreading? Were gulls out of this world for the Bible's translators, or were they keen to indict a more familiar (and more wicked) bird, the cuckoo?

Robert Alter's recent scholarly translation of *The Five Books of Moses* (2004) suggests that gulls were indeed among the bogey birds. He gives us a new go at chapter eleven of Leviticus:

> And these you shall abominate of the birds, they shall
> not be eaten, they are an abomination: the eagle and

the vulture and the black vulture, and the kite and the
buzzard according to its kind, and every raven according
to its kind, and the ostrich and the night hawk and the
seagull and the hawk according to its kind ...

There are hardly any gulls in Shakespeare. Edward
Armstrong – Ulsterman, clergyman, birdman – wrote in
Shakespeare's Imagination (1945), that 'there is no indication of
a personal acquaintance with any sea-bird ... Shakespeare had
not much knowledge of the sea.' *The Tempest* has Shakespeare's
only reference to gulls: Caliban, in his drunken wooing of
Trinculo, offers, to get him 'young sea-mells from the rock'.
Even this mention isn't cast iron: some editors print 'scamels'
and gloss the word as a textual corruption, or suggest it might
mean a goat, a shellfish or a 'godwit'. Other gulls (meaning
fools, and those who fool someone) crop up more commonly
in Shakespeare. The person duped is a gull; the trick itself was
also called a gull. Elizabethan England had many *gull-gropers*.

The bird's English name has an appropriately coastal history
coming out of Welsh, Cornish, Low Breton, Middle English,
Old Norse and Old Irish words for 'wail' or 'wailing' – from
the bird's sharp cries. The gulls that are dupes might get their
name originally from an Old Norse word for yellow (referring,
possibly, to the golden down of newly hatched birds of any
species). An appetite for swallowing all sorts of junk might also
come into gulling. Gulls do have gullets. Perhaps somewhere
the meanings might join. 'Gullible' survives in our vocabularies.
In Shakespeare, *Twelfth Night* has most gulls: Malvolio and Sir
Andrew Aguecheek are both called 'gulls' in the play. A 'gull-
catcher' in these terms is a sharp or cozener. From this family
tree, the OED lists 'gullery', 'guller' and 'gullage'.

In his collection of notes about the nature of Norfolk that became a manuscript of sorts in the 1660s, Thomas Browne subjects birds to a systematic order. His section on marine birds begins:

> A white large & strong billd fowle called a Ganet
> which seems to bee a greater sort of Larus. whereof I
> met one kild by a greyhound neere swaffam another
> in marshland while it fought & would not bee forced
> to take wing another intangled in an herring net wch
> taken alive was fed with herrings for a while it may be
> named Larus maior Leucophaeopterus as being white
> , & the top of the wings browne.

Here, on the page, as he pauses at his own name, we can follow Browne's thoughts, watch him thinking about the sea and its seabirds: first the gannet and then gulls. Skuas follow in his text: they are wilder birds than gulls, darker, dirtier, more errant, usually pelagic, habitually piratical. He knew of a single county record of the great skua: 'one whereof was shot at Hickling while 2 thereof were feeding on a dead horse'.

The skua pricked Browne's memory of gulls. 'The Larus maior [is seen] in great abundance in herring time about Yarmouth,' he says. Next, the big gull makes him think of the little one, the black-headed, and the birds have become part of a human story:

> Larus alba or puets in such plentie about Horsey that
> they sometimes bring them in carts to norwich & sell
> them at small rates. & the country people make use
> of their egges in puddings & otherwise. great plentie

thereof haue bred about scoulton meere, & from
thence sent to London.

I ate black-headed gulls' eggs once. A clutch of three, soft-boiled and toothsome, tasting of wet grass or a slightly high salad. A tart marsh. And the yolks, fogged sunrises loosed onto my plate, came sliding from the most beautiful homes I've ever cracked. Chalk-buffed by their boil, the shells looked like old washed flints.

Black-headed gulls, in England at least, were the first to challenge the family's oceanic reputation. In William Turner's terms, the black-headed wasn't quite a gull. But, he noted, one hundred years before Thomas Browne, the gull family's predilection for man-made cast-offs:

> There is a sea-bird, like a Daw in size, but with the wings
> sharper and longer, wholly white in colour, save for a
> black patch which it bears on the head, and with the beak
> and feet of purplish red. I often, journeying upon the sea,
> have had this bird in mind ... The anchor being weighed
> this bird immediately flies to us in the company of Gulls,
> promising something to itself by way of food out of the
> refuse cast forth from the ship; at last exhausted by its
> constant cries it merely utters 'keph', as Gulls cry 'cob'.

Black-headed gulls had been farmed for eggs and meat in gulleries in Britain since the thirteenth or fourteenth century. Wild colonies were owned and part-fenced. Gurney gives a list. In England in the seventeenth century, there were at least twenty working gulleries, including the ones Browne knew in Norfolk and several in Essex. Young birds were caught before they could fly, fattened in pens on bullock liver, and served at table. There was a risk that they tasted too much of where the family hailed from; the lean meat was said to have 'the raw gust of the sea' (Gurney quoting Thomas Fuller's *The Worthies of England*), but it could be sweetened if the young birds were further fed with 'gravel and curds, the one to scour, the other to fat them in a fortnight ... their flesh thus recruited is most delicious.' Larger gulls also carried the marine taint, but could be made palatable. Thomas Muffett, doctor-writer of Elizabethan England, is quoted by Gurney:

> Sea-mews and Sea-cobs feed upon garbage and fish [and are] thought therefore an unclean and bad meat; but being fatted ... they alter their ill nature, and become good.

Though some people ate some of them, all gull species remained little known to almost anyone who lived more than a mile or so from the sea. John Clare, nineteenth-century poet and expert birdman who roamed the fields and fens of eastern Northamptonshire, didn't know herring gulls. To him they were seabirds ('said to be common about Boston') and he never recorded the species in or around his parish of Helpston, a mere thirty miles from the saltwater Wash. Clare himself only saw the sea once in his life, and few gulls of any species feature in his poems. In one, set in the flooded fens, they appear as alien forms: 'strange birds like snow spots o'er the huzzing sea'.

Selborne in Hampshire is twenty miles from the open sea as a gull might fly. Gilbert White's *Natural History* (1789) doesn't mention any. And, in his *Journals*, White, the tireless early patchworker, struggles to place or name those he does see:

March 13th 1771: Woollmere Pond – Some large white fowls also: qu: what? They had black heads.

Three days later, he'd sorted them:

... it seems probable that the gulls that I saw were the pewit-gulls or black caps, the larus ridibundus. They haunt, it seems, inland pools, & sometimes breed on them.

Not long after Gilbert White's gull-noticing, Samuel Taylor Coleridge did some gull-*imagining*. Around the 1790s, before he'd begun to incubate his albatross, Coleridge composed a short poem that may stem from a Bristol Channel sighting of a herring gull but which is, as his biographer Richard Holmes has noted, 'an emblem of the poet at work, seen in different moods'. It is also, like all of Coleridge's birds, amazingly well *seen*.

Sea-ward, white gleaming thro' the busy scud
With arching Wings, the sea-mew o'er my head
Posts on, as bent on speed, now passaging
Edges the stiffer Breeze, now, yielding, drifts,
Now floats upon the air, and sends from far
A wildly-wailing Note.

★

I, too, have hurried out to watch the gulls in the Bristol Channel. My parents, in their seventies, moved from Bristol to Minehead. It was a mistake. They quickly became stuck in the seaside town with no easy means of getting anywhere else. We joked about *the last resort*. My father lost his nerve as a car driver and got scared of the hills of west Somerset. My mother suffered a period of illness that left her immobile and trapped indoors. They survived, but it was tough for a time. Dad could manage the level roads to the supermarket and went on little sorties, bringing back titbits to try to revive his permanent bride. She was confined to the front room, and he, like a bowerbird, decorated it with cheap CDs and discounted pastries. Sitting at her side, he leaned in to show her his billets-doux.

I went to help and to be as kind as I could, but by mid-afternoon on most visits I needed air. I took my binoculars and walked to the front. There's a bit of beach there, and a freshwater drain that opens below the sea wall and spreads across dark worn pebbles. Herring gulls and black-headed gulls wash in this stream, and I hoped that if I watched, I might pick up one of the rarer birds I was trying to learn about. I didn't. Once, I continued walking along the beach towards the seaward end of Minehead as the sun was setting and there, where the road stopped, on the steeper, rockier shore, was an adult Mediterranean gull in winter plumage. I knew it right away among the black-heads: the overall crispness of its whites and a black muzz about its head; its sour face and blood-coloured beak; its stompy feet and legs – a serious bird among clowns.

My parents had been responsible for my first Mediterranean gull at Oxwich. More than forty years ago, it was they who, one half-term, took my school friend, Richard, and me to the Gower in South Wales (across the silty sea from Minehead and west a bit). The gull we saw there was similarly plumaged to the Somerset bird, an adult, heavy and white and obvious. Mum and Dad waited by the car park on the beach as Richard and I noted down what we could see. In a cupboard at the back of the room where I am writing now I have kept the letter accepting our description.

These are no longer required. I saw the bird in Minehead again, the next time I went looking. It was in residence there through the winter. It may well be a British breeding bird. They weren't when I saw my first but they are now, with more than a thousand breeding pairs and large late-summer roosts. But the bird on the beach had enough of an escape artist about it for me, and I used it for that.

★

Black-headed gulls were the first urban gulls in Britain as well as the first regular inland gulls. In the mid 1920s, E. M. (Max) Nicholson (then in his twenties, later a champion ornithologist and conservationist) planned to write a popular book to be called *Bird-Watching in London*. It was never finished but has been recently printed. Gulls, specifically black-headed gulls, are at its heart. By this time, the species had joined the house sparrow and the wood pigeon as part of the urban bird flock. 'As a family,' Nicholson wrote, 'the gulls are among the most interesting British birds for in their status, and consequently their whole habit of life, they are among the least stable.' Gulls, he recognised, were evolutionarily dynamic ('lost tribes of the plover race', he says, that took to the sea) but also culturally adaptive, and moving in on the opportunities afforded by man-made environments. Most of Nicolson's observations are from the city's parks. The Thames (where gullers now pan for rarities and ring numbers) gets little attention. Some species hardly occurred then in London. He saw singles only of lesser and great black-backed gulls.

All the gulls Nicholson notes are seasonal visitors, and mostly appear in winter. The Round Pound in Kensington Gardens was then the place to see black-headed gulls. These birds could be studied closely. He notes how rain makes them 'truculent'; he watches one that is able to fly backwards and others that dive for grain and bacon-rind beneath the pond surface like 'little gannets of this little sea'; he observes them playing games with twigs and leaves on the surface of the pond; he studies piratical practices against sparrows and tufted ducks; he tracks, from the upper deck of buses, the gulls' flightlines over London to their roost sites, as they followed straight paths 'which they knew without Euclid to be the shortest distance between the two points'; he discovers in a single gull among the flock a pugnacious winter-feeding territoriality, 'a thing undreamt of in the philosophy of the territory theory'.

Nicholson also wisely deduced that wintering gulls in London were likely to come from various breeding populations: 'neither purely English nor all strangers but a cosmopolitan crowd of English, Scottish, Dutch, German, Swedish, Danish, Estonian and Lithuanian birds.'

Before Nicholson, the arrival of black-headed gulls in London was described by an immigrant bird lover W. H. Hudson (who'd started his birdwatching life in Argentina). His book, *Birds in London*, appeared in 1898. The previous year H. G. Wells's *The War of the Worlds* was serialised, and Bram Stoker's *Dracula* was published: there were a lot of aliens and invaders abroad, ngendering much cultural anxiety. The gulls first came to ondon in hard winter weather and, for the time Hudson knew 'hem, their numbers rose and fell, on the Thames and on ponds lakes in the parks, according to the creep and freeze of ice. v were refugees from the cold, and prompted a passing flurry of human interest among the sideshows and street life of the late Victorian city. In the winter of 1887–1888, gulls came upstream as far as Putney. Hudson quotes a Mr Tristram-Valentine:

> their appearance has, from its rarity, caused a corresponding excitement among Londoners, as is proved by the numbers of people that have crowded the bridges and embankments to watch their movements. To a considerable portion of these, no doubt, the marvellous flights and power of the wing of the gull came as an absolute revelation.

Some were keen to kill the birds and took guns to the river. Hudson was against this. In the winter of 1892–1893 there was a long and severe frost. The shooting of gulls on the Thames was stopped:

> And then for the first time, so far as I know, the custom of regularly feeding the gulls in London had its beginning. Every day for a period of three to four weeks hundreds of working men and boys would take advantage of the free hour at dinner time to visit the bridges and embankments, and give the scraps left from their meal to the birds. The sight of this midday crowd hurrying down to the waterside with welcome in their faces and food in their hands must have come 'as an absolute revelation' to the gulls.

The winter of 1894–1895 was equally harsh: there were 'hummocks of ice' on the Thames. Hudson himself went to the river with sprats. Nature, the gulls had taught, might be nurtured and a new culture was born out of a new urban entanglement with the residual wild. 'A tradition formed', Hudson observed. It is noticeable that he describes the charity of 'working' men. The gulls had something of the poor and hungry about them, and they were first recognised and aided by the class who would have known the human equivalent of gull need.

Hudson liked the gulls and admired the workers' generosity, but we can see here the beginnings of a motif or meme that has dogged gulls ever since. They are plebs: a hungry posse, downmarket incomers from elsewhere, rowdy in town, canny and opportunistic, physically strong and able to work hard, but, given the chance, keen on hand-outs and loafing. That word 'loafing' is still associated with gulls more than any other bird.

In 1926 the nature writer and *Times* correspondent Anthony Collett published *The Changing Face of England*, and a year later *The Heart of a Bird*. Like a number of 'country' writers who wrote between the wars (I've mentioned the Essex champion, James Wentworth Day, but there are others: H. J. Massingham, Henry Williamson, C. Henry Warren, Arthur Ransome), Collett wrote natural history as a kind of national history. Mostly the news was bad: the time was out of joint, the old ways dead, the ancient continuities severed. The writer's job therefore was to capture the dying truths. Help came from birds and other animals that might stand in for the paradise lost. Some, unfortunately, reminded you only of the Fall.

Collett was a keen observer and wrote about London's black-headed gulls coming for crusts on the Embankment and in St James Park, as 'an endless dance of moths with lantern eyes'. He liked the birds' flight; or rather he liked them where he thought they belonged, out of town. In their scramble for scraps in the city, there was something ugly and wrong:

In London [the] feats of flights are more stereotyped. They are adjusted to picking food from the scum-head of the flood-tide or to narrowing the return circle to the cornucopia of the Embankment office boy.

For all their mothy whites, the gulls had become dirty birds doing dirty things. They were egregious beggars whose panhandling revealed their gypsy qualities.

These gulls' shouts, whether of fear or greed, need the silence of wide spaces to mellow them to a sound like laughter; in London even the background of the grinding tramcars leaves them harsh, cold and defiant. They return the Londoner no affection but cupboard love.

As they wheel close past us to seize their dole in the air, always approaching up-wind, we gain a keen impression of the eager and remorseless gull nature. That ridged beak, that stretched throat and cold eye betray cruelty and greed, and though the black-headed gull is far less of a butcher than the larger kinds which sometimes visit London, he, too will snap up small fledglings in his summer haunts. See how his scarlet legs and bill set off the purity of his French grey back and white breast and tail, which he usually contrives to keep clean even in London fogs, or when he roosts upon a coal barge.

A lot is being loaded onto those wings. The shifty bird's success is fabricated from foreign feathers; its apparent cleanliness and purity of plumage has nothing to do with it being immaculate, but is a disguise contrived over time in order to duff and deceive; the birds are evil. It is hard to account for the vitriol here. Collett's more direct state-of-the-nation writing in *The Changing Face of England* might explain it. There, he associates the gulls with a wider malaise. The country risks rot by being coddled. Nature has been allowed to let rip and go large. Sanctuaries and protection have replaced the necessary control and subjugation provided by the boss man and his big gun. Left unchecked, indulged by *boys*, the gulls have mutated:

> It is clear that to continue the present policy of universal immunity in bird sanctuaries is merely to invite the robbers to kill off all other species. A change must come, and it is time that public sentiment was awakened to its necessity. Means must also be planned of reducing the numbers of gulls in sanctuaries without disturbing the shyer inmates. Taking the gulls' eggs is not enough to reduce them, and the use of a gun, except possibly some unusually powerful air-gun, would most likely disturb the gulls little, and drive away the always fickle and fugitive terns.

Even the victims in Collett's world get a kicking from the prosecuting counsel. You can feel him putting on a uniform as he writes this speech. But although he is armoured, it doesn't really help. Showing how troubled he is by the mixing of gulls and people, of wild places juxtaposed with the built world, adaptation and co-evolution making new mongrel categories of life, his ideas and rants bleed across species, and he laments debased human taxonomies and fears the urban hybrid:

> In London and other great cities our nation seems to be losing the fresh tints of a race nursed in the forests

and by the sea, and to be assimilating itself to the physical habit of cave-dwellers.

★

At the Coastal Park Landfill in Cape Town, no one stopped us at the gate. Claire, my wife, drove our battered old car right up to the tipping face, as if we were about to make an offering. We had climbed dozens of metres of compacted trash and were higher than anything for miles. Just inland from the Atlantic beach at False Bay, the landfill heaves over the Cape Flats. Seen from a distance, its sheared top, dirty and dusty, mimics the great bulk of Table Mountain on the northern horizon. The smoke and stream of gulls circling the tip rewrites the mountain's fabled tablecloth of cloud. From far away you can see how the new world answers ancient stones.

Close up, the dump is most alive at its face, a slow-breaking wave of fresh-dropped rubbish. The wave trembled where the dust-carts tipped. Men sifted skips by hand as clouds of dirt were added to bags of household waste. Flip-flops were items to look out for here; I counted four of them pushed into the creamed filth. At the face, the air was especially particulate, a hot buffet, thick with sun-heat, sea-salt, sand and grit. A compactor-macerator with metal-toothed wheels surfed down the wave front, rupturing whatever lay beneath it. And there were the gulls, like soldiers at the heel of a siege engine, huddling and dashing, hovering, walking, hovering again, moving as the machine did, entranced by its manufacturing magic, its *turnover*, and taking whatever it let them snatch.

Hartlaub's and kelp gulls worked the tip in their thousands, along with smaller number of grey-headed gulls: all three of South Africa's short list of regular species. With the gulls were sacred ibises, dirty shirted, and a troupe of goats, likewise colourless beyond grime. The gulls are strangely able to maintain a *sta-prest* appearance. They remain clean and look clinically white. At the fringes of the eruption a few less committed rubber-neckers took stock: kites wheeled overhead, cattle egrets stalked the out-field, and a pair of white-necked ravens stood, black-suited and unimpressed, looking like auditors of the lot.

A high sea in False Bay could breach the dump's dusty ramparts should it wish, but the dump might retaliate by bleeding its toxins into the saltwater. It dates from a time when the sea was a place to rubbish. Next door, and a product of the same mind-set, is Cape Town's sewage works. Next door but one is the Cape Flats, where decades ago the white rulers of South Africa decided that the blacks of the city should live. They still do, though much has changed since the fall of the apartheid regime. That said, none of the people we saw working on the dump were white.

The gulls are mostly white at Coastal Park. Adult plumages predominate and all of the gulls keep their whites. All the human officials we saw wore dust masks. None of the human tippers or those searching the waste did. The gulls were panning too: owning bags, stealing nuggets, hanging around the blunderbuss of an off-white pelican and filching whatever it had missed. Only twice in an hour did I see something I could identify as edible: a pie-crust and a naan. Yet at every moment some bird or another had picked up something that might pass as food. And so: shouts and fights and pursuits and thieving.

There was more fighting over food when, one week later, Claire and I went after gulls again. The terms were different, though the birds might have been the same. It would be only a short flight of a few miles across the Cape Peninsula from the dump to the Atlantic shore at Olifantsbos. Here, the city and its trash are hidden behind mountains, and the sea is in charge and something like *seagulls* are apparently at large.

We'd gone out on a wreck-walk following a map that marks the final destination of the many ships that, over the centuries, failed to round the Cape of Good Hope. Between long black hawsers of kelp branches, ripped by storms from the seabed and flung onto the beach, were various skeletons: rusting iron vessels and a small boat smashed into bleached spars of driftwood.

We turned a low headland, sand pebble-dashing our legs in the roaring wind, and I saw what I thought was a rust-coloured rock with six squabbling kelp gulls on it. Then, through the salted air and the rotting weed, we smelt the scene. Before we got too close for their comfort, three or four gulls dipped their burly heads to the caramel smoothness between their feet. Their beaks went inside it and ate or drank there. It was a humpback whale. Dead already and washed ashore, it showed its nakedness and indignation. Its rorqual pleats were now uppermost and ran down its length like a ploughed field. The gulls stood on one ridge or another and worked holes into the blubber.

The overturned carcass seemed deboned. We could make out flukes and a flipper, but nothing seemed calcium-solid. Below its fatty disintegration, there was no face, no skull, no ribcage. Who had taken that? Where had they put it? The whale might have been melting. And the gulls had arrived to stand on it and, doing so, seemed like some prehistoric sea vulture, older than all whales, than every iron wreck and broken wooden skiff, older than the attending undertaker mimed by the white-necked raven on the beach, older even than the sea itself.

Junk Bonds

Barney is eight years old in Clive King's book *Stig of the Dump*. He lives, not far from where I did at the same age, on the chalky downs of Kent. Out playing one day, he stumbles on a worked-out quarry when the ground gives way and he falls into a pit and meets Stig, who lives there. Stig's home is a dump within a dump; Barney is transfixed by a grotto furnished with rubbish:

> He'd never seen anything like the collection of bits
> and pieces, odds and ends, bric-a-brac and old brock,
> that this Stig creature had lying about his den. There
> were stones and bones, fossils and bottles, skins and
> tins, stacks of sticks and hanks of string. There were
> motor-car tyres and hats from old scarecrows, nuts
> and bolts and bobbles from brass bedsteads. There
> was a coal scuttle full of dead electric light bulbs and a
> basin with rusty screws and nails in it. There was a pile
> of bracken and newspapers that looked as if it were
> used for a bed. The place looked as if had never been
> given a tidy-up.
>   'I wish I lived here,' said Barney.

Barney isn't sure if it is still 'Nowadays' down in the dump or how old Stig might be. 'About eight? Eighty? Eight hundred? Eight thousand?' But the pair find a way to communicate and their adventures begin.

Stig does the noble savage very well; he is a Home Counties cave man but classless and socially pure, awkward with no one, and backward enough for his lifestyle to be ahead of everyone's. There are repeated good jokes in the story about who is civilised and who is smart. At a fancy dress party near the end of the book, Barney goes as Stig but Stig arrives as Stig too. Who will play the part best? It could be a Shakespearean comedy, something that happens at midsummer or on twelfth night, in a forest or in Bohemia. And in this way Stig joins a pedigree: the unlettered and pre-modern *dumpstars*, knights of the open road, with holes in their shoes or perhaps no shoes at all: the Wild Boy, the Leech Gatherer, Steptoe and Son, Vladimir and Estragon, Huck Finn lighting out for the territory before he gets *sivilised*, various supertramps and poachers and young Heathcliffs, all those that grunt truth to power and central heating. Everyone would like a dustman like Stig camped just over the garden fence, a friendly hominid living off roadkill and recycling all our crap. Except that now we have the homeless and food banks and refugees and millions below the poverty line doing that, and it has all got a lot *messier*.

The book teaches this in its own way. It is striking how clean the rubbish seems in the story; it is pre-plastic junk, non-toxic, a wholesome *raw material* rather than irreducible waste. But it is also interesting that the best replica of our first domestic habitat is a dump, that Stig's cave is littered and he's at home there, living among rubbish. As dumps have been neighbours to us – little more than chucking distance away – ever since we took up settled existence, so litter might define us as apes on the way to today. Society has had to be disposable and throwaway to become modern.

When they had the Great Plains to roam, the Crow Indians left their campsites on the Little Big Horn without a scrap of trash. Their enemies would have tracked them had they left any evidence behind. Now they are corralled on a reservation in Montana with nowhere to ride but the road out, and their refuse builds around them, hobbling their ponies' hooves and flapping on barbed wire like desperate flags.

There is a horizon-expanding end to *Stig of the Dump*. Stig takes Barney and his sister Lou back into the past to something like the beginning of England. They witness a huge stone circle being erected by hundreds of Stigs with great revelry on Midsummer Day. Afterwards, the three get home into the same day of their present and time jogs on. Stig had led them back, but then he disappears. Where has he gone?

> Actually the dump's filling up fast now, and Stig may be on the move. One report was that he'd been seen working at a garage by the main road, where they collect old wrecked cars and put the pieces in rusty piles. And somebody else said he saw him in a back lane of that woody country at the top of the Downs mending a chicken-run with an old wire mattress. It certainly sounded like Barney's friend Stig, but perhaps it was only a relative of his.

★

Before *Stig of the Dump* there was Samuel Beckett. Among his works that make particular play with junk are *Endgame* (first performed in 1957) and *Happy Days* (first performed in 1961). Dust is required on stage in both. In *Endgame,* Nagg and Nell, the parents of Hamm, live in two ashbins. In *Happy Days* Winnie is half-buried on a dumpy mound.

Clov, Hamm's helpmeet and adopted son, has the opening lines of the play. They might concern his blind master, who is sleeping in an armchair covered in an old sheet, or they might float a vision of the wider world:

> Finished, it's finished, nearly finished, it must be nearly finished. [*Pause*] Grain upon grain, one by one, and one day, suddenly, there's a heap, a little heap, the impossible heap.

We're all up to our waists, or deeper still, in our remains. And we're all waiting to join the dust. The living residue is what Beckett is about. He writes from near the exit point. Many, if not all, of his people are preoccupied with de-existence, but none of them can quite yet move or leave or die. The waste of themselves remains. As does the actual rubbish of a lifetime, memories as well as objects, our *self storage*: lint in a trouser pocket, an old hat, a bag, and countless thoughts, the stuff we cannot shake off, our world without end.

One of Clov's tasks in *Endgame* is to turn a telescope on this world 'without' and to birdwatch: 'Do you want me to look at this muck heap, yes or no?' he asks. Yes, Hamm has need of his sightings. Clov reports limited grey views and not even the given grey bird of such a scene. 'No gulls?' asks blind Hamm. 'Gulls!' replies Clov, meaning there are none, meaning it is ridiculous to assume any givens, meaning that Hamm's memory of a sea scene cannot be answered by anything other than 'GRREY!' – a world greyer even than gulls.

My wheelie bin has *No Hot Ashes* stencilled on its lid, though I have never lit a fire in any house I've lived in. Nagg and Nell live in bins that are like urns for living ashes. As with many of Beckett's cast they are not quite dead. Meanwhile, they are embarrassing parents. Nagg tells rude jokes. They feed like toothless babies on 'pap', which could be milky porridge or could be a milky breast. The oldies' presence, popping up out of their bins, raises questions. Are children the waste of parents or are parents what children must discard in order to live fully? Hamm wants his father, his 'accursed progenitor', over and done with, finished and removed, like shit fetched away as nightsoil:

> My kingdom for a nightman! Clear away this muck!
> Chuck it in the sea!

Hamm seeks his own way out too: 'Put me in my coffin.' But, as Clov says, there can be no homes for the dead: 'There are no more coffins.' There can be no final clearing away, no decluttering or *life laundry*. It might all come to rags and bones but there is no absolute end of any story. Ashes give way to ashes and dust to dust.

The play is in mourning for death, for a time when things could die, when life went round and began again, where nothing lasted forever. Both Hamm and Clov turn out to be depressed ecologists and moribund romantics. Clov: 'I say to myself that the earth is extinguished, though I never saw it lit.' They both half-remember when they could see things themselves or when there was something to look at down a telescope, when there was colour other than grey in the world, and when there wasn't just ash and dust. Hamm has green dreams: 'If I could sleep I might make love. I'd go into the woods. My eyes would see . . . the sky, the earth. I'd run, run, they wouldn't catch me.' And again: 'here we're down in a hole. [*Pause*] But beyond the hills? Eh? Perhaps it's still green. Eh? [*Pause*] Flora! Pomona! [*Ecstatically*] Ceres!' He's pastorally susceptible but he knows it's not going to happen. He tells a story in which he rebukes a beggar who has called on him: 'But what in God's name do you imagine? That the earth will awake in spring? That the rivers and seas will run with fish again? That there's manna in heaven still for imbeciles like you?'

There is no green; there aren't even any gulls.

The same goes for *Happy Days*. The setting seems to be a beach; there is sand and sun, though no avian signifiers. Winnie and Willie might be on holiday but are really stuck in some hellish waiting room. Ways of leaving life or the slow business of getting into the ground are concerns throughout. Winnie has made a start. She begins, buried to her waist, by declaring that it is a happy day. In the second act she is buried up to her neck. The mound is a burrow for her but also a great gown or robe tumbling from her shoulders. She's dressy and has a bag to match. Much of the business of the play has her organising her stuff. She does her face. She opens a parasol. She talks to a red revolver – 'You again!'

Willie, hidden behind the mound, stirs intermittently and reads out choice items from the newspaper. There's a saucy postcard. The sun is hot. The parasol catches on fire. At one point Winnie directs Willie into a shady hole so that, quoting the dust-dirge from Shakespeare's *Cymbeline* ('Golden lads and girls all must, / As chimney-sweepers, come to dust.') he will 'fear no more the heat of the sun'. It could just be an old couple mumbling over breakfast, letting slip the emotional catastrophe of their lives between brushing their hair and browsing the news. Winnie talks unstoppably and nervously; Willie is grumpily monosyllabic. It could just be a domestic crack-up, a marriage losing its mind and petrifying. But Winnie is half-buried in a pit, and all her witter only underlines her crisis and its question: how soon might it be possible to die?

> If I were not held – [*gestures*] – in this way, I would simply float up into the blue.

And a little later:

> Shall I myself not melt perhaps in the end, or burn,
> oh I do not mean necessarily burst into flames, no, just
> little by little be charred to a black cinder all this –
> [*ample gesture of arms*] – visible flesh.

Ashes, however, are not granted to all.

In the second act Winnie sinks lower. Willie appears briefly but says even less than before. Only Winnie's head is visible now. It is a head full of rubbish, the junkyard of a life. She has rare memories of happiness but otherwise it has been bleak: a lost childhood, missed dates, failed meetings, people staring at her if she were in a zoo. Is Winnie going mad? Or has she become a sort of world-transmitter, a mouth on a mound, an anti-oracle speaking not of what is to come but of what has passed, a chatty archive of everything that is unendingly sad?

★

In Bristol, I walked from my desk to buy a sandwich. As I began to cross a side street, a herring gull dropped from above and landed ten feet away in the middle of the road. It had spotted a stray chip on the tarmac. The gull landed on fleshy feet, its wing-limbs folding away like two silver-weapons sheathed, its beak hard as bone, its eyes bright and icy. Looking both awkward and brazen, it was an animal come to town. It walked as I did, with two legs, but was otherwise from altogether somewhere else. A car drove past, keeping me to the gutter and flushing the gull. With a big flap it pulled up from the road between the walls of the buildings. It was climbing into the air. The car cleared the chip and I crossed the road for my lunch. When I came back the chip was gone.

Needs

The only gull on the British list that James Hanlon hasn't seen is a slender-billed gull. A *mega* for twitchers – only eight of them have been recorded in Britain. The last was a one-day bird at Titchwell in Norfolk in May 2014. It stayed for less than an hour. James went for it but dipped. 'I'm not really a larophile though,' he said to me. 'But there'll never be an end, there'll always be something I need, something I haven't seen.'

We were at Grafham Water, and larophile or not, James had offered me gull lessons and some gull talk. And gulls are good birds to be keen on in the flatlands of the East Midlands/West Cambridgeshire area. Some landfills still yield birds, even nowadays with reduced or negligible food waste deliveries. And there are reservoirs like Grafham and the winter floods of the Ouse and Nene Washes, where great numbers of gulls continue to come to roost. When James moved with his wife and children to the outskirts of Cambridge, he followed the birds: 'I started doing gulls in the winter because Milton tip was close.' He's been watching them for more than ten years, more seriously in the last four or five: 'You take what you can get.'

I'd seen his reports online and his Twitter posts and I'd read his book *UK500: Birding in the Fast Lane*, which describes his quest to see 500 species in Britain and Ireland before he turned thirty at the end of 2004. I would have adored this book when I was fifteen and a neophyte twitcher. At fifty, I was still impressed. To my surprise, I was jealous too.

James is a civil servant working on UK immigration. I asked him whether he thought there was a connection between chasing after rare bird visitors to Britain and working with human migrants. He didn't think so.

He grew up in London. Aged six he was keen on dinosaurs, aged seven he was 'noticing birds', by eleven he was regularly out birdwatching, and he was twitching from thirteen. Later he spent eighteen months, on and off, accompanying Lee Evans, the first publicly known hardcore twitcher. That would itself have been an education, but mostly, James said, he learned his birds 'by looking at them.'

We met on a November afternoon, as the gulls began to assemble to roost at Grafham. As we stood in a wet grassy field at the edge of the calm reservoir, lines of birds arrived from all around us. At first they were serious white against sombre grey and then, when the sun slid low, each was transformed into a bright metallic flake. The big numbers arrived close to dusk and settled on the water only when the sun had gone. There are just a few minutes to scan and process these birds before the dark takes them out of sight. Their calm meets the birder's hurry.

James is sharp. I rambled on about my gull confusions as I looked through my binoculars, but he worked the gathering flock of herring and lesser black-backed gulls. He picked up a winter yellow-legged gull on the water. This was the first I had ever seen in Britain. It was obvious once he'd pointed it out – it had a strikingly pale head among the much darker young herring gulls. Once I'd seen it, I could relocate it if I looked away but, soon after, I had the sensation that I used to have in maths lessons in school: leaving the classroom, my grasp of the subject slipped and blurred with every step I took away from it.

Before it got too murky we saw one more striking bird: a petite and dark lesser black-backed gull. It was longer winged, blacker, and noticeably smaller than the other birds; it could have been a Baltic lesser black-back. I was excited; I'd read about this subspecies, and this was an intense and handsome bird. James was not so hopeful. They are very hard to prove, he said. There were overlap problems with the *intermedius* subspecies of lesser black-backed, among other things. Most ages of the bird would still need a ringing recovery to be acceptable and the one in front of us wasn't going to get anywhere.

Later that winter, we went out again; James offered me another roost-watch at Fen Drayton, a series of flooded gravel workings, near Cambridge. There were fewer birds than we hoped but James talked more.

'Gulls are a challenge,' he said.

'Since I was about nine I have been trying to identify every bird I see. There are always ones that get away but it's also good to try to develop your instinct to follow through on birds you don't clinch.

'You have to watch gulls for years. Any flock is bewildering. Their aging processes are hugely complicated. You often have to scratch your head. I don't like them particularly as birds. Some look downright ugly especially when they are moulting, but they don't deserve their bad press and for me they present all sorts of interesting questions.'

We waited for some. The sun was starting its exit. Mist began to creep from the edges of the pit.

'The herring gull just a few years ago had yellow-legged gull as a subspecies. People used to pick them out occasionally in late summer in Britain. Then some time in the 1980s the yellow-legged was split from the herring gull, and it was then split itself with the Caspian gull emerging as a species. It is confusing and for some that's a reason not to get into gulls. The i.d. is difficult and the taxonomy is in a state of flux.'

There was nothing doing in front of us. I asked James to rehearse some identification pointers. He said this was the sort of thing he used to do in his head when he was driving to a 'rare'.

'For Caspian gulls some of the best features are structural, they have long legs, long wings, a long slender bill and a small head; they are also white-headed and dark on the back like yellow-legged gulls. In juvenile or first-winter plumage they have a heavy shawl of grey steaks on the neck that contrast with their white unmarked head. Little clues give it away; if you watch gulls regularly they stand out a mile.

'All gull watchers like Caspians. They are regular now across the Midlands and East Anglia. If you work your local landfill you'll find one in the winter. They didn't even exist when I started and to have seen the rise of them is special. That said, I do think it is pretty arbitrary where you draw the line as to what is a species.'

This remark surprised me. You say that, I asked him, even with a biology degree and all your interest in sorting all these birds out?

'Well, my degree was a long time ago and you'd be on a fool's errand to argue about species. That is best left to the taxonomic committees. The revolution with gulls has been mainly about splitting species. The idea of a species seems a requisite for us because of our need for order but I think nature doesn't quite work like that, it is all fluid really, or maybe a fairy tale.'

Have we reached a point where the taxonomists are outstripping the birders' field skills? Are some declared species now inseparable except by DNA analysis?

'It's already happening. There are lots of cryptic species emerging. And that is not useful for birdwatchers who have to rely on what they can see and hear. And again, these gulls do hybridise: Caspian and herring, herring and lesser black-backed.

'One January, I saw a strange gull right here. It looked like an Iceland gull but was odd; it didn't have their white wing tips. I decided it was a race of Iceland, a Kumlien's gull, which are said to be the result of a *hybrid swarm* [a phenotypically mixed population resulting from multi-generation interbreeding] between Iceland and Thayer's gull. Kumlien's breed on Baffin Island. There are only six or seven records of Kumlien's in the county. I put out the news. The bird stayed but was hard to see and I began to have some doubts. But others saw it and we convinced ourselves. Gulls are a nightmare.'

Since we spoke, the adjudicating rarities committee returned this bird to James as 'not proven' as a Kumlien's gull; they decided he hadn't eliminated the possibility that the bird was a pale Thayer's gull. By that time James had other doubts of his own. Some gulls cannot be known.

'But to find a first would be like winning the lottery. Once in a while you'll find a rarity, once or twice in your life it will be something astounding. Killian Mullarney, a great birder, found the first Vega gull for the Western Palaearctic in Ireland in 2016. It's the eastern Asian counterpart of the herring gull and a very subtle bird, it looks similar to herring, but a combination of features, a dark eye, a dark mantle, the blotching on its head and neck – and crucially it being in active primary moult – secured the record. It is a big story. I went for it but missed it. Black-tailed gull and kelp gull are due now, they are on the radar, and who knows what else.'

There were still no gulls where we are. It was getting chilly. James swapped his baseball cap for a knitted beanie. Before he put it on he rubbed his hands in it to warm them.

'I am interested in all wildlife. I've seen all the British butterflies and the bats. I am a mammal lister as well as a birder. I arranged an expedition to see the last remaining black rat colony in the British Isles, on the Shiants off the Outer Hebrides. When I heard these rats were being exterminated to protect the seabirds I rounded up some people and we went. We chartered a boat and stayed on the islands and trapped some and watched others coming to bait. It was a new mammal for us all. Like half of our mammal fauna they are not native but they have been here a long time, they were introduced a thousand years ago and spread from ships and, yes, they played a major role in the spread of plague, but I see them as a British mammal and was sorry to hear they're being exterminated and was really pleased to see them.'

There'd been a desultory show at Fen Drayton. The gulls must have gathered elsewhere. As the light sank everything began to fade in front of us. James had one last story from the Hebrides. I asked him if pursuing vagrants ever felt to him, as it did once to me, like a kind of coffin-bothering, a blasted bird hounded to death's door, and all that. James had thought about this.

'We don't know what happens to many vagrants. It isn't nice to see a sick one but I don't feel I am chasing ambulances. The reality of bird migration is that huge numbers of birds set out and many perish and we cannot do anything about it. It isn't different for rarities. Most, in any case, will likely survive.

'I desperately wanted to go for the white-throated needletail that turned up on Lewis in 2013. It is a near mythical bird for twitchers, one of the most wanted. After a mad dash I got up there the next day and saw it: the fastest bird in the world with a level flight speed of 100 mph. It was an amazing thing to see, I was spellbound, watching it zooming over the moorland. We sent the taxi driver we'd booked back into town to get us fish and chips, he brought them back, and we scoffed them on the verge, and we watched the bird, and as it flew past a wind turbine it appeared to collide with the nacelle and suddenly fall, limply, to earth. We couldn't believe it. I found it on the moor beneath the turbine. It had been killed outright. The fastest bird in the world, taken out. The first in Britain for more than twenty years. From the Pacific to the Outer Hebrides. And then that. Surreal.'

★

There are still gulls to see at Milton landfill just north of Cambridge. Clay pits dug there in the 1970s have been receiving waste since the 1980s. Parts of the tip are now capped with soil and are described as 'restored'. They look like low tumuli against the flat fen sky, new burial grounds ghosting old ones. Some areas of the tip are still operational and currently 96,000 tonnes of waste arrive each year. Mostly this is inedible, but some gulls still find things to eat.

Through much of the first decade of our new century, Dick Newell watched and photographed and thought about the gulls at Milton and other sites in Cambridgeshire. His attention brought much of the mess of the family, its new species and subspecies and hybrids, to other birdwatchers' notice. His austere website – grey screens of minimal text surrounding hundreds of photographs from Milton and elsewhere – has been visited nearly half a million times.

I had planned to go to Milton on my own to see what I could see. The site was celebrated for Caspian gulls. Then the news broke that the tip was being searched for a man, Corrie McKeague, who went missing in September 2016 after a night out in Bury St Edmunds. He might have taken shelter and fallen asleep in a commercial dustbin and been inadvertently picked up and then killed by the crushing action of a dust-cart. Bury's waste goes west to Milton and perhaps Corrie was tipped there. There was a twenty-week search of the dump for human remains, but nothing was found.

I hadn't gone before and couldn't face going while the waste was being sifted for a body or whatever might be left of a body. I'd seen gulls do their own version of that at Pitsea. I'd spent my twitching youth chasing wind-wrecked transatlantic vagrants and Siberian blow-ins. I'd read my Henry Mayhew on London's waste workers and had been out at night on the Thames with the body-salvagers of Dickens's *Our Mutual Friend*. I stayed away from Milton. My telescope wouldn't have been welcomed by anyone and I don't think I could have used it. The hunt for the body resumed in the late autumn of 2017 in a part of the landfill adjacent to the area already examined. After seven fruitless weeks the search was called off.

Jonathan Livingston Seagull

To study the cultural presence of gulls, the uses we have put them to, is to come upon strikingly opposing representations. Gulls have been identified with purity and with danger, with sublimity and with filth. Like pigeons, we've granted them a contrary status in our bestiaries, where they have become useful projections for our ambiguities about who we are. Gulls and pigeons are both holy and hated. The winged rat can be the homecoming friend; a dove can be a peace emblem or a stand-in for the Holy Ghost. The gull can be a consummate flyer or a predatory glutton, an instructor of grace or a depraved hooligan.

I hated *Jonathan Livingston Seagull* for years before I read it. It arrived as a bestseller as I was getting serious about birds. Published in the USA in 1970, it first appeared in the UK in 1972, and invented the feel-good/win-out/learn-from-nature book. Thousands, perhaps millions, were sold.

In 1972 I was an eleven-year-old ornithological zealot. I saw enough of the little blue book to know it was poisonous. It used the nonce word, *seagull*; it gave a bird a name that wasn't its common or scientific one; it attributed speech and human thought to its subjects; worst of all, it plastered the whole with a meaning quite other than what gulls surely would countenance.

In many ways I feel now as I did then. *Jonathan Livingston Seagull* robbed the gulls of America of their own truth. It stole their *gullitude* and colonised or domesticated the birds even as it constructed a fable around their real aerial mastery.

Both its author and its photographer were flyers. Richard Bach had been a US Air Force pilot, and was, the blurb tells us, 'now seldom without an airplane'; Russell Munson owned a Piper Super Cub, 'from which he took some of the pictures in this book'. Both looked hard and jealously at the gulls they described and photographed. As a jump-jet enthusiast might, Bach has a good description of a gull beating its wings to stall or halt in the air and then dropping down lightly to land on a beach. But, beyond the airmen's envy, their looking is put to cod ends.

The book begins with the young Jonathan practising his flying; he's already singular and a loner, 'no ordinary bird'.

For most gulls, it is not flying that matters, but eating.
For this gull, though, it was not eating that mattered
but flight.

His mother urges him to be like the flock, but he is 'off by himself', learning about speed and getting good at flying faster than the other gulls. But trying to set 'a world speed record for seagulls' he crashes 'into a brick-hard sea'. This crisis sends him back to the flock and he tries to rejoin it and be 'a normal gull'. But he cannot deny his deeper drive; he must be true to his gift, and off he flies again.

Already Jonathan is far from bird. He flies upside down, 'the first aerobatics of any seagull on Earth'; he reaches terminal velocity – faster and faster to 214 mph; 'the speed was power, and the speed was joy, and the speed was pure beauty'. Bach uses his knowledge of machine-powered flight to write of daredevil stunts and thrills: 'He discovered the loop, the slow roll, the point roll, the gull bunt, the pinwheel'.

But the elders, representing a state and its laws, do not approve. Jonathan breaks away again, and, on his own, learns how to sleep in the air, how to reject the 'stale bread' of fishing boats by riding high winds inland 'to dine on delicate insects'. An outcast now, he finds a new height, a new realm, a kind of burial in a 'perfect dark sky'.

The book's second part begins on this astral plane, where everything has collapsed into light. There is no need to eat; sex has never been mentioned. Apart from Jonathan's mother, there are no female gulls.

His feathers glowed brilliant white now, and his wings were smooth and perfect as sheets of polished silver.

That sounds like an aeroplane. By the time *Jonathan Livingston Seagull* was written, aeronautics was supersonic and deep into the jet age. Those magnificent men in their flying machines were no more. No one except a few hillbilly barnstormers were turning loops or flying upside down. High-altitude planes were bombing Vietnam, Laos and Cambodia without seeing their targets; Operation Rolling Thunder and other missions dispatched more bombs to these small countries than were dropped in the entire Second World War.

Jonathan wants to go fast, but in many ways his book is an elegy for the old days and the old ways of flying. That fits more comfortably with the book's emerging philosophy than charter-flights and city-hops, or carpet-bombing and helicopter gunships. Much of the second part of *Jonathan Livingston Seagull* is about mastering skills and self-control through meditative concentration. The flying manual turns out to be about what's inside your head.

Jonathan can already use 'an easy telepathy' or thought-transference instead of 'screes and gracks'. Then he meets a guru gull called Chiang who advertises a blissed-out nirvana: 'you will begin to touch heaven, Jonathan, in the moment that you touch perfect speed'. You can be everywhere and nowhere, Chiang says. You might get high as a kite and taste the euphoria of hypoxia, well known to free-falling pilots. But 'you must begin by knowing you have already arrived'.

> The trick, according to Chiang, was for Jonathan to stop seeing himself as trapped inside a limited body that had a forty-two inch wingspan and performance that could be plotted on a chart. The trick was to know that his true nature lived, as perfect as an unwritten number, everywhere at once across space and time.

Jonathan gets it: 'I *am* the perfect, unlimited gull!' (This, briefly, weirdly, has suggestive taxonomical implications as an assertion of the now refuted ring-concept of herring gull speciation around the Holarctic, which made for a kind of planetary crown of potentially unlimited gull taxa.) But the guru has an even more mind-blowing offer: 'We can start working with time if you wish,' Chiang says, 'till you can fly the past and the future.'

Around now the book re-orders its systematics, swapping the pantheism of the open air for a more Christianised hierarchical theology. 'Each of us is in truth an idea of the Great Gull, an unlimited idea of freedom.'

Jonathan returns to the flock and recruits some new generation rebels, Fletcher Lynd Seagull and Henry Calvin Gull, beach boys badged-up like California Mormons. The disciples and their messiah prepare for a crusade, a righteous battle flight. It's an air-show; the jocks finger their joysticks, 'while the Flock huddled miserably on the ground'. Here comes a 'wizard of low speed' Charles Roland Gull, and Martin William Seagull, a stuntman, and Terrence Lowell Gull, and Kirk Maynard Gull:

> eight of them in a double-diamond formation,
> wingtips almost overlapping. They came across the
> Flock's Council Beach at a hundred thirty-five miles
> per hour, Jonathan in the lead . . . The squawks and
> grockles of everyday life in the Flock were cut off as
> though the formation were a giant knife . . .

A butch-Christ has shown his stuff. After this trouncing a more humble and healing mood prevails, with shades of a last supper conversation. Future plans look floaty but they are gilded with the manifest destiny of all high-flying Americans.

> He spoke of very simple things – that it is right for a gull
> to fly, that freedom is the very nature of his being, that
> whatever stands against that freedom must be set aside,
> be it ritual or superstition or limitation in any form.

Some say Jonathan is a devil; some say he's 'The Son of the Great Gull!' We've heard his story before.

★

In the mid 1970s when my parents had a copy of *Jonathan Livingston Seagull* it sat on their shelves, next to what I naively took to be an answer to it, Erica Jong's *Fear of Flying*.

Walt Whitman's poem 'Crossing Brooklyn Ferry', a nineteenth-century idea-of-America text, might be a better work to put alongside *Jonathan Livingston Seagull*. It is as singular – a subjective passionate vision – and it too has gulls. They are not Jonathan gulls. The mix and blur of the flying birds 'oscillating their bodies' as the ferry crosses the East River from Manhattan are suggestive to Whitman; they become part of the great polymorphous mass of peoples and things and places and times that he traffics. W. H. Hudson watching the workingmen of London feeding winter black-headed gulls would have understood Whitman.

I too many and many a time cross'd the river of old,
Watched the Twelfth-month sea-gulls, saw them high
in the air floating with motionless wings, oscillating
their bodies,
Saw how the glistening yellow lit up parts of their
bodies and left the rest in strong shadow,
Saw the slow-wheeling circles and the gradual edging
toward the south,
Saw the reflection of the summer sky in the water…

*

In the end, I bought my own copy of *Jonathan Livingston Seagull*. American herring gulls had been split as separate species from herring gulls in Europe, and they began to be seen or claimed by gullers on this side of the Atlantic. I found myself wondering if the grainy black-and-white photographs of marine gulls in *Jonathan* might help me get some purchase on this bird. Also, I was curious to identify Jonathan himself.

The book feels very west coast, but are its gulls? American herring gulls are found on the shores of both oceans. The Pacific coast also has other large, potentially confusing species: California, glaucous-winged and western gulls. What was Jonathan? Wondering whether anyone anywhere in the world had ever done such a thing, I found myself poring over the black-and-white Jonathan photographs and holding them up alongside the colour pictures in Klaus Malling Olsen's gull identification guide.

In *Jonathan Livingston Seagull* there are some strikingly dark chocolaty young birds. They suggest American herring. The adults are trickier to identify. I wrote to Russell Munson and asked if he remembered what gulls he had photographed and where. He kindly replied.

> Most of the birds in *Jonathan* are [American] herring gulls, with some great black-backs as well. About half of the pictures in the book were taken in Rockport, MA, in 1963–1964 while I was a teaching fellow in photography at Phillips Academy in Andover, MA. This was years before I had met Richard Bach, and before the book was written. I just felt compelled to take pictures of seagulls. The remainder of the pictures in the book were taken in 1969 on Eastern Long Island between East Hampton and Montauk Point after reading the manuscript and seeing that more pictures were needed to better illustrate the story. As a matter of fact, I still love photographing gulls and do so to this day on Eastern Long Island.

I've never seen an American herring gull in Britain, though I've been near two when gulling in eastern England. One first-winter bird was seen in the pig fields at Great Livermere in Suffolk in April 2017. Before that, in February 2016, a juvenile was reported feeding in a 'chip skip' at a McCain's potato-processing factory at Whittlesey in Cambridgeshire. This record – a rare gull, a 'new' species, making its way in the Anthropocene by stealing chips from a dump – might in a single moment have floodlit everything that I am in pursuit of here. I wish I had seen it.

Into It

I went again to Grafham Water in west Cambridgeshire, this time to meet Mark Ward. It was a late summer evening, roosting gulls were our concern. I'd met Mark before: we'd coincided on a September trip to Fair Isle, where with Ade Cooper he found a North American buff-bellied pipit, the island's first. Ade and Mark were a friendly pair and tolerated me tagging around with them for a bit. Both are high-class observers.

Ade was working then as a carpet-fitter and Mark had a job with the RSPB. By the time we met up at Grafham he was editing their magazine, *Nature's Home*. He lives only a few minutes from the reservoir and goes often to watch its gulls.

'It can be worked in nicely with your significant others.'

The evening was still. The sea, wherever it was, felt very far away. The sky looked oily and the water below ran thick; insects, pollen and moulted feathers crumbed across its surface. There were bronze, silver and then gold arrivals of gulls as we spoke.

'The gulls can roost anywhere on the water,' Mark said. 'We know they're coming but we don't know where they'll settle. I get a good evening about one in five. If it was winter, I'd be straight in there looking for the white wings of glaucous or

Iceland gulls. And if I can pick out a white-winger, I feel like my job is done. Today, it's yellow-legged gulls and Caspians that I am after. I haven't had a Caspian yet this half of the year. My challenge is to get an August one, the first wave of fresh juveniles is here in the country but I haven't had one yet, but anytime now.

'I haven't met people who say, I *quite* like gulls. You're either into it or you're not. It's not a great wildlife experience for many people but it's my type of birding. Gulls and geese are my two favourite groups. Not many people like either but it's not about aesthetics for me. I like the calls and the sight of them arriving, but it's more the lure of finding something unusual and of getting through the i.d. challenges.

'In the winter, there's a race against time and the thrill of beating the dusk. In the summer, they're still coming in long after dark and you have to admit defeat at a certain point. It's very different watching on a tip. There are no legs visible here, there are no rings to read, but for numbers it is quite exceptional. The lesser black-backed record count for the county was set here, and I came within one hundred of it last year. I click count them in. It's good to have a target. I'd like to beat that count this autumn. The yellow-legged record was here too. I was pleased to set that one.'

Mark saw, as he was talking, the first yellow-legged gulls settling on the water.

'There are some yellow-leggeds coming in now with the lesser black-backs, six or seven of them, and just a couple of herring gulls. For me here the standard gull is the lesser black-backed. That's what I am comparing everything else to, especially since they're now breeding locally. You can find them year round, but some lesser black-backeds are still migratory and you get the yellow-leggeds with them, coming into the country in late summer, moving through it. They tend to associate with them more. It's great looking out tonight and seeing so many lessers.

'Yellow-legged gulls easily outnumber herrings at Grafham in July and August. There were fifty last time I was here, and I've had up to 140. The herrings come later and build up from the end of September onwards, the British ones first then the bigger Scandinavians. A day comes and they're all in.

'Here's some more just arrived. They're building now, three just dropped in. A nice adult preening. At this range both yellow-legged and Caspian juveniles are tricky. The young yellow-legged will be advanced in their moult. Adults are not so bad, even at a distance: they're paler grey than lesser black-backed gull but darker than herring, less cold, less blue, more grey, with a chunky head and plenty of black on the wingtips They look longer winged, too, though not as long as Caspians. They're distant but that's pretty good for here. There'll be lots we'll have to let go, but we're well into double figures.

'I admit it's an acquired taste. Lots of people would prefer other birds. I don't blame them, but I want to be here. I really believe the gulls are interesting. There's so much to learn, so much to think about. Look at the sky now! Full of gulls as far as we can see – I like that. They're always there for me in this part of the world. And tomorrow will be different.'

Within a few days of our meeting, a report popped up on BirdGuides.com of a least six Caspian gulls including a juvenile at Grafham Water. I asked Mark if it was him? 'It was,' he replied 'and there were yellow-legged gulls too, it was one of those rare *fill-your-boots* sessions.'

090518 – dormer projection –
gutters alongside stacks.

staggered roofs – covered pots

Talk to the Animals

In July 2015 – the summer silly season, when young gulls *wheedle* on rooftops and daft stories fill the news – David Cameron, then Prime Minister, called for a 'big conversation' about how something commonplace had stepped out of line.

Earlier that week some large gulls had attacked a tortoise called Stig in Liskeard, Cornwall. 'They turned him over and were pecking at him,' said Jan Byrne. 'We were devastated.' Twice in the previous three months, it was reported, gulls had killed small dogs elsewhere in the south-west of England.

When pets are attacked, the nation is shaken from its summery doze – shaken enough to buy a paper at least. It was funny when you saw them shit on your mate, it was cheeky when they pinched your ice cream, but this...

Chris Johnston took down Cameron's words on BBC Radio Cornwall for the *Guardian*:

> I think a big conversation needs to happen about this, and frankly the people we need to listen to are people who really understand this issue in Cornwall, and the potential effects it is having. Reading the papers this morning about how aggressive the seagulls are now in St Ives, for instance, we do have a problem.

In Liskeard, Jan Byrne, who had lost Stig, was worried that the gulls would come back for George, her other tortoise. He hadn't been his usual self since they'd got Stig. 'After the attack,' she said, 'George was extremely subdued.' Petal, Jan's rabbit, was traumatised too: 'I think she must have seen the whole thing.'

In Newquay, Emily Vincent was mourning Roo, her Yorkshire terrier, who was badly injured in a garden gull raid and couldn't be saved. It was especially galling that the killers, herring gulls that nest on local roofs, were protected. The gulls left a 'murder scene' Emily said. 'He had crawled back into the house and collapsed . . . He was on his side in a pool of blood.' Worse still, her three-year-old child witnessed the carnage.

In a Honiton garden in May, a chihuahua puppy had been killed by gulls. This warm-up act for the summer allowed Robin Page, writing for *The Daily Mail*, to reheat a dead duck. Conservationists and legal protection, Page wrote, were to blame for nature getting 'out of hand' and tipping the balance in favour of birds of prey. The swelling numbers of predators were being allowed to kill all that they wanted. British sparrowhawks (the 'supreme killers') were enjoying a meat feast that Page translated as 24 million blackbirds or 88.2 million sparrows a year. His argument was a mess (the RSPB estimate the total winter population of blackbirds in the UK to be between ten and fifteen million birds, and that there are under six million pairs of house sparrows); and his descriptions straight out of Alfred Hitchcock:

> Her little Chihuahua had been playing in the garden for only a few minutes when the seagulls swooped down and pecked it to death. Nicki Wayne, 57, was

having a shower when her puppy Bella sneaked out of the door into the garden of her home in Honiton, Devon. Nicki knew of the dangers. She had seen the seagulls watching Bella on previous occasions and so made sure she was with the dog whenever it went outside. But this time she wasn't there.

The gull panic was printing itself from here on.

Some must have been minting it, too. But to keep the story earning through the season it would need pepping up. First they came for our chips, next for our pets, and then . . .

Steven Morris in *The Guardian* on 23 July spotted it:

The latest attacks in Cornwall, south-west England – which appears to be the hotspot for attacks this summer – have left a 66-year-old woman needing hospital treatment and a four-year-old boy traumatised after his finger was savaged. Sue Atkinson was pounced on as she walked her dog in Helston. She said: 'All of a sudden I felt this whoosh. I saw it was a seagull and he came in again and cut the top of my head. I couldn't see what was happening – I was oozing blood. It did frighten me. Apart from the fact I was bleeding, I was scared it was going to come back. It was like a scene from the film *The Birds*.' She staggered home and was taken to hospital, where her wound was treated.

Four-year-old James Bryce was attacked as he munched a sausage roll during a holiday in St Ives. The gull appears to have been guilty of bad aim rather than evil intent, going for the snack but getting

James instead. His father, Alex, said the incident had left James 'petrified' of seagulls. He said: 'We've been coming to Cornwall for years and seagulls have always been a problem, especially in St Ives, but things have definitely got worse. Five years ago, if a seagull swooped for food, it would have been a 'wow moment' – now it's happening every five minutes and is more malicious.'

Two days later in *The Daily Star* the story effectively ate itself. Its lead on its front page (beneath an ad for free ice lollies) was all upper-case, SEAGULL TERROR: LOCK UP YOUR BABIES. 'Expert warns giant birds WILL kill tots' it said next to a photograph of a great black-backed gull wolfing a rat. A cull of the 'winged monsters' was called for: 'The Gull Awareness Group has demanded that the Government must take the problem of rogue gulls seriously before someone is killed.' The expert, the leader of the group, Simon Prentis, 'warned that it will not be long before a child dies'.

Two days further on and the same *Star* finished off the festivities. The paper hybridised two canards, a long runner and a newcomer. The story, now demoted, ran on page eight, with a sub-headline on page one, 'Seagulls mug dole spongers':

Psycho seagulls have a new target – Britain's benefits claimants. The giant dive-bombing birds have been swooping on residents in the UK's most poverty-stricken seaside town, Jaywick in Essex.

The *Star* had watched the TV and was borrowing from a show called 'Benefits by the Sea' which featured 'JP, a recovering alcoholic and drug user … cowering in fear on a walkway as a huge seagull prepares to strike':

> He tells his pregnant former heroin addict girlfriend Sarah: 'You know I don't like seagulls. There is one up there on top of a lamp-post. Them seagulls will come down and take the f***ing food out of our hands.'

★

Gulls and mass panics have coincided before. The birds' flocking behaviour was crucial to the development of one man's ideas about what, in the 1930s, he believed was thought-transference.

Every day for decades Edmund Selous watched birds. He was a pioneer field ornithologist and an acute observer. Notes written on the spot were his hallmark. They fill his books. He was interested in all bird behaviour but especially in courtship and in the evolution of social habits in colonial or gregarious species like ruffs, rooks and black-headed gulls. He transcribed and published his notes, often adding paragraphs written 'at home' (sometimes 'in bed') reflecting on the meaning of what he thought he was seeing.

He saw much that no one had noticed before, including, in 1902, what Tim Birkhead et alia call the 'bizarre precopulatory display of the dunnock', and in 1906 the way reeves – female ruffs – choose their partners rather than being chosen, and in 1907 how female black grouse mate with dominant males after leks. Selous was proving that Darwin's idea about sexual selection actually occurred in nature. His observations were original and triggered new understanding – or should have, had they been more widely known. Known or not, his thinking about his later obsession, thought-transference, was fascinatingly wrong. Less wrong than Jonathan Livingston Seagull's 'easy telepathy', but still off-kilter.

Like most Victorian naturalists, Selous started as a hunter and specimen collector but he turned away from bloodletting and what he called 'zoological morgues', and sought ways to find value in nature observation alone. His older brother, Frederick, had a celebrated career as a hunter and wrote books about his big-game escapades in Africa. He was killed in 1917 in a marginal skirmish of the First World War in what is now Tanzania. Scouting through binoculars for his enemy, he was shot in the face by an expert German rifleman.

With a more nervous and less slaughterous disposition, Edmund had already taken his binoculars to the tamer fields of the British Isles, where he watched and recorded and wondered. That was his credo. His observational diaries are extraordinarily detailed. But, looking so long and so hard, he saw almost too much, and his accounts are often too dense to be intelligible. Every move of a watched flock was annotated. But, a map of life the same scale as what it maps is no use. Worse still, he over-thought his findings. Trying to close in on the avian minds of those he was studying, he imagined the intricacies and intimacies of flocking birds. Thought-transference or telepathy was a pseudoscientific human craze in the late nineteenth century, and Selous believed birds were doing it as well. Perhaps it particularly appealed to his reclusive nature – you could be part of a crowd without having to talk to anyone.

Selous's thinking is hard to describe. His writing, especially in *Thought Transference (or What?) in Birds*, shifts between vivid specificity and hypnotic dreaminess; at times it feels like a conversation between a therapist and a patient; at times you catch the movement of his pen in his notebook as he hurries to get down what he's seen, then you can see him lift his head as he allows his mind to wander after what he's watched. Sometimes we might be reading the anonymous, small-type, double-columned pages of the behaviour section of the *Handbook of the Birds of the Western Palaearctic*, at others Virginia Woolf's *The Waves* – published the same year as Selous – with its mazy and meandering thoughts. This is part of an entry about black-headed gulls at Weymouth Swannery in Dorset on 31 October 1928:

> At once the band on the water rise, and as I turn
> my eyes towards the others, they meet the whole
> multitude, both of them and the peewits also in
> the air. All, though thus separated, must have risen
> simultaneously, and this appears to me to be an
> impossibility without the interpenetration of some
> pervasive energy through the entire multitude, which
> must have numbered many hundreds.

Selous is excited by what he has seen. The entry continues:

> Then some fly off individually – a dozen or twenty
> perhaps – to the entire indifference of the rest, the
> great packed multitude. But now – again see the
> difference – this latter absolutely at the same instant
> of time goes up like one enormous bird, which, in that
> instant, I believe, is very much like what it really is. It
> was a wonderful thing to see, but still wonderful is it,
> I think, to reflect that such things can go on by day by

day, through the centuries, either (for the most part) unobserved, or, if observed, unwondered at.

He wonders on. And, two days later reports 'from bed':

Alarm being excluded, what other cause through the normal channel of the senses can produce these great instanto-simultaneo upgoings of a multitude of birds distributed over a considerable space and not all visible to one another. The last circumstance seems to exclude the agency of sight. Outside themselves, however, there was nothing but the landscape, in this case an abandoned, waterlogged marsh without either human beings or livestock upon it. There remains hearing only, for I do not perceive how smell, taste or touch can come in. Well it was too dusky now for ordinary shooting, and the air had been unwounded in that way for the last hour perhaps. Other offences such as aeroplanes, fog-horns, sirens, etc., were equally absent. In short, the act and the circumstances together necessitate a cause outside the normal.

Unwondered at birds and the unwounded air – Selous's coinages are telling.

His moment-by-moment recreations of phenomena observed make a book that is gripping but also mad. An almost-genius is channelling the life of black-headed gulls and coming as close to sitting directly behind their dark eyes as any human could. He is like a programmer writing code for them. But is there anything madder than the index entry he compiles?

Black-headed Gulls: anticipatory tremulous movement in band of, before simultaneous rising; successive simultaneous flights of; origin of gregarious instinct through thought-transference, suggested by actions of; interplay of thought-transference and independent action shown by; numbers of rising altogether, as 'the collective gull'; pauses between successive flights off, of, best explained through thought-transference; continuous miracle witnessed in flight of; collective movements of, of three kinds; difference between flocking and feeding-flight movements, of; twenty-five rising as though joined together; alternating changes in character of flight, of, through collective and individual impulse; simultaneous rising, through thought-transference, of separated gatherings, of; one gathering, of, not otherwise put up by rising of another; vast gathering of, rising like one enormous bird; such simultaneous risings part of normal habits, of; simultaneous risings with almost immediate realightments of; successive simultaneous risings of, unaccountable except through thought-transference; great scientific demonstration given, by; process of transfusion of thought-wave, amongst, observed; two modes of departure, of; actions of forty-five best explained by thought-transference…

The next index entry reads: 'Blake (William): Was on right track when drawing the ghost of a flea.'

It is easy to laugh, but we might also love Selous for his detail and his passion, and for his reaching out towards the uncapturable. And was he so far off? We know now that flocks function when individual birds take account of a few of their neighbours, and that such local reading of nearby conditions spreads through the mass so that thousands can appear to move as one. I have often seen the ripple of the collective gull, and the dread that can infect a flock. Sometimes it is just a question of terms. Under *Dread*, it says in the *Dictionary of Birds* (1985), see *Panic*. Each age gets the hysteria it deserves. It suits us today that gulls are piratical chip thieves invading our spaces. But the same birds were once configured as hyperborean strays from far out at sea. The concept of thought-transference emerged as investigations into crowd psychology developed. Ideas move through the human mind: last year's hysteria becomes this year's shell shock.

Selous saw as much but called it by a different name. Unfortunately it was a name that suggested too much thought. Nor was his naming straightforward. He wrote too singularly about the action of a collective. He avoided contact with other bird workers. His books had odd, questioning or tentative titles. They were all 'unreadably' put (David Elliston Allen) in a 'florid' style (Tim Birkhead et alia). Little known, they were also not widely reviewed by the ornithological establishment. Had more people noticed, his early original findings might not have given way to later projections or fantasies. Selous was an ethologist without knowing it himself, and few early fellow animal behaviourists (Julian Huxley was an exception) knew of him. Although his reputation grew posthumously, when he died in 1934 there was not a single obituary.

*

The only bird I have ever seen at the moment of its natural death was a black-headed gull. On my way to Chew Valley

Lake, I watched one fall dead from the sky onto a road in south Bristol. It was part of a flock heading from their overnight roost on the reservoir back towards the city where I was coming from. Its falling flailing wings caught my eye as I drove beneath it. The flock opened slightly around it, just enough to let it fall through them as they flew on. None looked anywhere other than where they were going. Dead already, the gull spun down to the edge of the road where it immediately looked like any other dead bird. I didn't stop.

Fly Tip

Dominic Mitchell met me at Rainham Marshes RSPB reserve and drove us to the shore of the Thames at Havering. As he drove, he told me about his other car. He'd sold some unwanted bird books and raised £600, he said, which he put towards buying an old Jeep. Like Paul Roper at Pitsea in Essex, this vehicle lives on a landfill site and will, at some point, die there. Dominic uses it as a mobile hide – he's sat in the Jeep observing gulls for many hours and, cumulatively, for weeks and even months. Before, he had to be escorted onto the tip by the site staff and was left to stand in one place. He was often very cold and wet.

Dominic is the founder and managing editor of *Birdwatch* magazine, and also runs BirdGuides.com. With him at Havering Park, I hoped for a Caspian. Having fed on the dumps nearby (Rainham, Pitsea) large gulls like to bathe in the Thames then preen and loaf on rocks there. It was the Essex landfills along the north shore of the river where this gull was first identified and raised into life as a British bird. As we pulled into a parking place a handful of milk-chocolate immature gulls were taking to the air in front of us from a wet patch of rough ground, and among them Dominic spotted a second-winter yellow-legged gull. We scrambled out of the car. The interrupted ordinariness of the moment tripped me up, as it always does. No amount of bumping into birds prepares you for it. Birds do this all the time, but gulls do it in places where those of us who are not primed are barely looking, in corners or edges of the built-up world, where the texture of the scene doesn't commonly include birds as large and *present* as gulls. Suddenly there they are: lined up on the roof of a warehouse, landing on a station platform, stalking a pavement next to a waste bin, working a puddle in a car park, calling on us throughout our dreary homeland.

We walked downriver along the bank of the Thames, keeping our telescopes and tripods slung low to the ground since Dominic had noticed that the gulls don't like the extra legs. We talked, when panning and scanning, in the kind of meditative way that you do on a car journey at night. To the south of us, plane after plane headed west. High over us others went north. Gulls were moving all the time. The tide turned as we followed the brown river's edge; our talk's rhythm and flow shaped and driven by the birds.

'Though lesser black-backed gulls are very common here, herring gulls are the default large white-headed gulls.'

At Grafham Water in the summer, Mark Ward uses lesser black-backs, here Dominic sees *through* herrings.

'The European herring gull itself shows huge variability. You have two subspecies here: *argenteus*, the taxon that breeds in Britain, and *argentatus*, the Scandinavian herring gull, which is larger and a more robust-looking bird. *Argentatus* are regular here in large numbers in winter: the adults are slightly darker grey above; the juveniles moult later and often keep that youthful plumage longer than our *argenteus*.'

This is one of the splits that must be done by the guller, a carving at a subspecific level.

'To be honest, because of individual variation and the difference in size between males and females, it is not always easy to say, clinically about every bird, which one is which subspecies. But classic *argenteus* is smaller and the adults are paler mantled, and if you can see them side by side there is generally an obvious difference.'

And here my vertigo began again, on the river's edge, close to sea level. This is why, when away from books and pictures and my gull friends, I can still look at herring gulls and think, *oh god*.

'It takes a large gull basically four years to go from a chick to an adult. It moults every year, its feathers are bleached by the sun, there's variation within every species and there are often other very similar species – when you're presented with a flock there's every possible variation. But they are so accessible too. I live in suburban north London; I can be among gulls within minutes of my home. They come to us. They exploit where we live, town parks, tips, and the river here. I've just been to Morocco for my wedding anniversary, but I managed a bit of gulling and found a glaucous gull, the first Arctic-breeding gull I've seen in Africa.

'You can see why it is such an attractive place to gulls here. They're using the Thames as a flight corridor from the sea and this is a good place for them to feed and gather. The tip at Rainham has been here for several decades, as long as I remember. Some of east London's rubbish used to go out further downstream, to Mucking, and that tip (it's closed now and remodelled as Thurrock Park) featured in a lot of the first Caspian gull records. The BOU [the British Ornithologists' Union] took their time to split yellow-legged gull (from herring) and then Caspian (from yellow-legged). But there's no one now who doesn't regard them as separate species.'

I like the idea that observers in the field were noticing differences in these birds before either the geneticists or the taxonomists woke up. Dominic himself has compiled a new annotated checklist of the *Birds of Europe, North Africa and the Middle East*. No one has produced such a list for twenty years. Birders are increasingly keeping Western Palaearctic lists, not just British ones, and Dominic's list names 1,148 species ready to tick.

'A sunny day is not the best for gulling. I prefer it cloudy, because when you have the sun glaring on white birds with pale grey backs it becomes much harder to judge certain features like their mantle shade.'

I asked Dominic about the gullers' *fifty shades*, the Kodak Grey Scale, a standardised colour grading where one is white and twenty is jet-black.

'Even if you don't have a copy of the chart itself, it's a relative measure, so if a bird has a mantle shade of Kodak Grey Scale three and another of four and a half, you know the difference is going to be detectable by eye, and you can call up the grey in your mind.'

The Scale is used to help gull greys be meaningful throughout the two contemporary doorstops of gulling: Klaus Malling Olsen's *Gulls of Europe, Asia and North America* (illustrated by Hans Larsson) and Steve N. G. Howell and Jon L. Dunn's *Gulls of the Americas*. It would be hard to overstress the importance of Malling Olsen's book to British gullers. Little more than a decade ago, it and the advent of cheap and good digital photography changed the way that it was possible to look at gulls in Britain.

Dominic, like my other gulling friends, carries Malling Olsen's 600 double-columned pages in his head and has managed to marry it with his optical acuity. Me – it makes swoon. Probably the densest bird book I know, it is almost beautiful and almost mad. It reads like dreamy grit, arresting my attention and dissolving it at the same time.

'A few Caspian gulls winter here and others move through, but there are never that many. The species is probably relatively new here, but it has spread west, starting somewhere in western Asia, and now breeds in central Europe as close as Germany. After they've bred, they move further west, dispersing along river systems and come out into the North Sea, and a small number make it to Britain.

'A few years ago the first British record of Caspian gull changed: it is no longer a bird from Mucking in Essex in 1995, as retrospectively a photo of one in Dorset in the mid-1980s has been accepted. So it's a fact that it's several decades since the bird first came to Britain. The last winter waterbird census estimated between 90 and 100 across the country but I'm sure that's too low. Some sites have up to ten on a good day.

'The first I went to see was a well-known bird that used to winter at Woolwich ferry in east London. It was commonly regarded as a Caspian gull but since it has gone, and in the absence of good pictures, the bird has been questioned. I don't know whether it would pass muster by today's standards.

'Sorry to interrupt,' Dominic was panning his telescope along the river as he spoke. 'There's a gull flying with a blue bag caught on its leg, one of the perils of feeding on the rubbish tip. It's also showing a lot of white in its wing tip. I'd like to get to grips with that. Look at its outer primary, it looks all white at the tip; it could just be a herring gull, but a classic adult Caspian often has an all-white tip to its outermost tenth primary. It's shed the bag, now and it's landing. I'll try to zoom in, have a look. No, that's just a herring gull. Even though its wings are closed now, the jizz said herring right away.'

Between Dominic's 'look' and his 'no' were three seconds only, little more than a breath.

'The thing about gulls and watching gulls is that sometimes there are no answers and that's always challenging. You must be happy to say you don't know. You can find internationally renowned ornithologists disagreeing on identifications, and that is the challenge of gulls. They are cryptic. But, unlike the crossbills or the reed warblers, they are around us all the time. So it's a challenge anyone can have a go at.

'People who aren't into birds think they're the same birds every day but they're not, of course. That yellow-legged gull we just saw hatched last year either in France or Spain or Portugal. It's here today, but maybe later this spring and certainly next spring, if it survives, it will be back in the colony it started life in. It doesn't breed here but the species is regular here. This unprepossessing landscape is part of its life, and it works for these gulls.'

The gulls on the river continued their toilet. Think of the *cloaca maxima*, one of the world's earliest sewage systems, and the Thames as a poisonous sewer, a highway of shit; but think of it redeemed, cleaned even, by the richness of gull traffic capitalising on its flow.

'You could come here on many bleak winter days and see hardly anything part from gulls, but its great to understand how gulls view this place, and I've been on top of that dump and watched them arriving from as far as the eye can see, from beyond the Dartford Crossing, coming in, dropping down, feeding, resting and moving on in, regarding this place as a focal point in the landscape. In midwinter half these birds may have come from Scandinavia, some from the Arctic Circle. Unless you spend time sifting and trawling you'll never appreciate the diversity or understand what would bring a bird here from the Arctic or anywhere else.

'Here on the dump, in January 2015, I had an Iceland gull with a ring. I saw it well. The bird was ringed two years before in Norway as an adult. But it would have bred in Greenland, hatched on a western cliff, then drifted to Scandinavia, where it was caught; then it spent the following year in Denmark, where it was found by people I follow on Twitter, and then this year it turned up on my local patch, only the third-ever foreign-ringed Iceland gull seen in Britain (the other two were sixty years ago in Scotland). And occasionally there'll be an even rarer bird; I found a Kumlien's gull here; they breed in the north-east of arctic Canada; and then early in 2011, the slaty-backed gull.'

The *mega*.

'That was a bizarre day, I won't forget it. I'd bought a new camera and wanted to try it out. But as I came down the A13, I could see this area behind us above the dump was absolutely black with gulls, swarming with them. No one can turn down a flock of gulls like that so I diverted from my original plan and came down here. I spent an hour looking over the fence going through all the gulls. I didn't see much of note, just one or two yellow-legged gulls, and I was close to giving up. I thought I'd scan just one more time and I went through a line of gulls – and suddenly there it was. I instantly recognised this bird as a species I'd seen previously in north-east Russia and Korea. I thought, Christ, slaty-backed gull!

'I seemed to be looking at a bird that was from the Pacific Ocean, which had never been seen in Britain before and only once in Europe. I quickly fired off some photos and then I had to ring people, I had to get other people to come to see this, so I rang the *Birdwatch* office where I work and then the information officer at the Rainham reserve centre down the road. As I was talking to him about it I saw the bird fly and lost it – I didn't see it again that day.

'I got back that evening and was still pretty convinced that that was what I'd seen. A couple of keen birders had come that afternoon and brought a book. When we checked the images they were actually quite a bit blacker than the bird I'd seen; that was the only time I had a twinge of doubt. But I sent photos out that evening to various gull experts and all the replies I got were pretty positive, so I posted details online and the next day about thirty people came along, some coming at dawn.

'It was a really miserable, filthy day. Raining continually. I sat on a large pile of mud amongst the rubbish from half-eight in the morning to three in the afternoon. The gulls were behaving strangely all day. I was sitting there feeling a bit sorry for myself when a gull flew in and landed about thirty or forty metres in front of me, and it was *the* bird – absolutely extraordinary because this is a place where maybe 15,000 gulls were wintering. I fired off some shots and got open wing pictures which are what you need most of all for almost all tricky gull identifications. It flew off but came back and the gulls finally began to settle. I pinned it down in a flock that was roosting and rang people – in the end everybody who was there managed to see it and the general consensus was that it was indeed a slaty-backed gull, the first one in Britain.

'The next day there were somewhere in the region of 1,300 people, stretched out like a scene from *Zulu Dawn* from one side of horizon to the other, waiting for the bird. Unfortunately, it didn't show. There was no mass sighting that day. It was seen again, but very erratically and it became a hard bird to pin down. From start to finish, it stayed six weeks in the Thames Estuary. It was seen here and at Pitsea, along the river, and at Hanningfield Reservoir, so it wintered in this area. But it was real needle-in-haystack stuff.

'We know it can turn up now, but the question is why? The answer seems to be the melting of the polar ice cap, which has made a new ice-free route around the top of the world. For the first time, there's an open water channel in the Arctic, and if you visualise the world then the distance between the Arctic Pacific and the Arctic Atlantic is actually quite small. Most birds will not wander between the two, they know their home range, but occasionally one might stray.

'Now there've been eight records in Europe, all adults or sub-adults. Young ones are much harder to identify and we probably don't have enough experience yet to do this over here. You need to have a good handle on what they look like and feel like in the field.'

'What's next?' I asked.

'There are several candidates: Vega gull, which is essentially the Asian herring gull, but they are hard work. Black-tailed gull, another Asian bird has occurred across North America as a vagrant and could turn up here. More realistic, though, might be the rarest of all birds on the British list that's not extinct: a great black-headed gull, also called Pallas's gull. There's only one record of this: a bird caught on a fishing boat off Devon in 1869.'

Dominic peered through his binoculars as he talked.

'I'm looking hard but still don't see a Caspian. Ah, but that's an interesting herring gull, to the right of the power station chimney – a really blotchy dark bird pecking under its left wing. It's worth a photo for sure.'

I struggled to find the bird that Dominic was looking at and, then when I did, I couldn't see the details he described to me. The difficulties made me anxious, but Dominic was relaxed. The birds, as he worked them, made him so.

'The name of the game is persistent effort. It's very hard to quantify what it is about gulls: you have to own up to an amount of obsession. It's a very repetitive kind of birding, a building of a knowledge base, finding things that are slight variations from the norm, that's what maintains the interest.'

Our talk about being interrupted by a bird was interrupted by another. 'Oh look at the primaries on that, just come down near us – a youngish bird, a herring gull type. I need a picture of that, I think it's a young *argentatus*, look how big it is, and it's got a weird primary pattern. These young *argentatus* often moult late and they don't acquire first-winter plumage as early as most other large gulls. This one still looks rather young, with very strong patterning, an interesting head shape, and it's very large, appreciably bigger, with a frosty look. The primaries hanging down below the tertials have odd patterns with pale areas. You don't see that often on a bird of this age.

'It's nice when you've got time with gulls. The line-up will change as the day goes on, and there's always something coming through. Do me a favour, lift up your tripod and hold it high above your head.'

My impromptu scarecrow-windmill worked and frightened the gulls nearby into flight. Dominic took pictures.

'Another problem bird. Every visit there are one or two to research afterwards.'

We walked back to Dominic's car. The river shone pacifically. The planes descending above it on their approaches were flaming in the sun. High up a gull, lit like a flare, was flying east.

'Look at that one right up there, far higher than the rest, flying straight through, heading out to the estuary. Some do this – maybe it is migrating. They don't all come down, some are like they're in a different world.'

Chayka

' I like them because they annoy us.'

Steve, known as Giggsy, was cutting my hair. Herring gulls had been swirling low to the street outside his barbershop in Bristol and we'd got talking about them. I knew he was keen on the neighbourhood wildlife. He'd broken off a previous trim to open his back window and coax a squirrel to a peanut and he told me of his plans to do the same with jays when they came through the city in the autumn. We hadn't spoken of gulls before. I suggested he might feed my hair cut to the birds.

'At home if I look up and see one passing, I'll throw something down and then watch them clock it. They turn, call once and then descend. I go inside, get my children to the window, open the curtains and there they are; it's like a circus in your back garden. The woman across the road here, in the chemist's, she hates them. I put bread on the top of the phone box and she came storming in here. They shit on her car.'

He had recruited gulls before, it turned out.

He remembered that they had twice shat on a girlfriend as they were splitting up, once in her hair and once in her eye.

'And I love that – hearing them on the roof, like they're wearing boots. In fact, there's nothing I like doing more than going out on a summer midnight in a caravan park to softly throw slices of bread onto caravan roofs, knowing that at four in the morning the gulls will descend with their clogs on.'

★

I saw Chekhov's seagull once: an adult Caspian that brassily kept charge of the low wall at the sea's edge in front of the writer's white-plastered cottage at Gurzuf. When he was in Crimea, Chekhov lived mostly in a rather stolid villa in Yalta, set back some way from the Black Sea. But twelve miles east along the coast, in what was then a little fishing village, he also had a more rudimentary bolt-hole, a beach hut on the rocky shore. I went a few years ago, before the Russians had annexed the whole peninsula, and I spent the last half-hour of a May afternoon looking at the gabardine green and tideless sea from Chekhov's terrace. Redstarts sang their sweet thin music from trees in the village behind me and gulls, passing along the slightest of swells offshore, got singed by the naked sun listing west. The bird on the wall probably had something going, scraps and all, with the keeper of the cottage, who didn't otherwise have much to do. At the Yalta villa there was a very different mood; the day before I was there, a Japanese visitor, declaring himself a scholar of Chekhov, had asked if he could hold the writer's toothbrush. He wasn't supervised adequately and somehow the relic was pocketed and removed.

Chekhov's *Seagull* is set inland; the gulls in the play come to a lake rather than living at sea. That wouldn't stop them from being Caspians – they breed on steppe pools and along rivers, far from saltwater, through southern Ukraine and Russia and east beyond the Caspian and Aral seas. But a more likely candidate for Chekhov's *chayka* is the black-headed gull. They are common inland and breed east across Russia to the northern Pacific coast. They also have a flighty and frail appearance and behaviour that would suit their symbolic appropriation better than any head-banging Caspian. That said, a kind of head banging is what is at stake in the play, the way minds mess with minds. The seagulls in it are required to represent drifting and unsettled souls. Nina, the would-be actress and lover of the would-be playwright Konstantin, declares herself a seagull drawn to him and his home, as the bird is to a lake. She is an emotional refugee, heat-seeking the blood of love and families, but she is never able to settle or rest in this world, and ultimately reflects only its falsehood and chill.

Early on in this demolition derby, Nina performs in Konstantin's debut play. It is an end-of-the-world monologue set 200,000 years in the future on a kind of rubbish dump. The play within a play is ludicrously bad and the beginning of the end for both writer and actor (the opening performance of *The Seagull* in St Petersburg in 1896 was also infamously disastrous), it is also a proleptic dream of Samuel Beckett's ashbins and heaps of dirt, in *Endgame* and *Happy Days*.

In Chekhov, we can watch a gull on the dusty tip. This is Nina reciting Konstantin's lines (as translated by Tom Stoppard):

No more do the cranes wake and cry in the meadows,
no more are the may-bugs heard in the lime groves.
There is nothing but the cold – the cold, cold
emptiness – emptiness and more emptiness – terrible
it is – terrible – it is terrible...
    Pause.
The bodies of all creatures that ever lived are as dust
– their indestructible matter is become stones, water,

clouds – and their souls have become one soul, and that soul is – me!

This girl-gull moves off and the next one we see is dead, carried on stage by its killer, Konstantin. His play has failed, his mother, a veteran actress, vain and frightful, has laughed at him, whilst Nina has shifted her attention to the more prosaic but much more successful writer, Trigorin. Konstantin lays the slain bird at Nina's feet, saying he will soon kill himself in the same way.

Nina is baffled. 'Look at this seagull,' she says, 'a symbol if ever I saw one, but of *what*, I'm sorry, I've no idea.' It is probably her lightest and brightest moment in the play.

Trigorin, watching from the wings, sees a story in the gull and takes notes when he enters to stand next to it and its avatar. Nina asks him what he's writing. He answers:

> Idea for story – young girl, like you, brought up on the shores of a lake. She loves the lake like a seagull and is happy and free just like a seagull. Then a man happens to come along, he sees her, and having nothing much to do, destroys her, like this seagull.

Two years have passed when the final act opens. Despite those early signs, the destroyer of the Nina-gull might not have been Konstantin. Instead, it is Trigorin who has stolen her life. Nina had a baby with him but the child died and Trigorin grew tired of her. Her acting became increasingly stilted. She signed herself 'The Seagull' in her letters to Konstantin – who still loved her – and eventually she returned to the home she did not have. Banned from her father's house, she is forced to stay in a hotel and refuses to see anyone until, like many a wind-blown vagrant on a stormy night, she taps at a window, as ghosts and gulls are wont to do. And Konstantin opens it, and opens up to her too, telling her of his love, his loss, his new writing success that still feels like failure, and his cold solitude since Nina left previously. She won't stay this time, either. She is distracted, a little mad. She leaves, not cured perhaps, but not entirely at her wits' end. But her arrival has tipped Konstantin towards his. He shoots himself just as the rest of the household bustles into the room – just as Trigorin declines to remember that he had, all those years before, asked to have the dead gull stuffed.

★

An aside after the curtain falls. I need to report an oddity, and this is the place to do it. As I worked on this book, I occasionally used the dictation device on my laptop. I speak the words out loud and they materialise on the screen. I enunciate as best I can but the ear of my Mac struggles with *gull*, and sometimes gives me *doll* or *girl*. 'To have the dead girl stuffed' – that took a rewrite. The errors can be funny, but they can be uncanny too, and I have shivered to see the alternative text and the suggestions that it has floated.

Gulls, gulls, gulls, as the pop band Sailor used to sing, didn't they? All the nice gulls like a sailor. T. S. Eliot traded in these terms too: there are 'sea-girls' that trip you up in the last lines of 'The Love Song of J. Alfred Prufrock'.

090518
vents dumm
our 'wire'
reap whites
3

White Gods

B ristol Temple Meads station, platform 15. I was early for my train and was taking pictures of a bold lesser black-backed gull that repeatedly landed close to people's feet and moved in on them, doing some freaking and casting its fluence, before lifting off, circling, and alighting again. A woman nearby and dressed in purple train-crew clothes, came up to me holding out her phone: 'I got him,' she said. Her photo showed a herring gull – not the current marauder – sitting on a junked toilet seat which had somehow found its way to the top of the low wall next to the tracks. The gull was ringed by the plastic seat and could have been warming eggs or taking a dump. The woman's name was Lesley and she sent me her photo.

*

'I'm a freshwater ecologist, I work on insects and fish. Professionally I have nothing to do with birds. I made a deliberate decision to keep my birding separate, as I didn't want to lose my hobby interest in birds or take the gloss off it. It has mostly been a good decision.'

Chris Gibbins spends much of his birding time watching gulls. I'd heard of his devotion to the birds of the coast of north-east Scotland, and I knew of his expertise having tried to learn from some of the articles and blog posts his gulling had prompted. He found an afternoon to talk to me in his office at Aberdeen University.

'My interest in birds got going at secondary school in Sunderland. My parents didn't have a car so we never explored the countryside, but a geography teacher, Mr Bates, ran a bird club, and once a month hired a coach and took us on trips. I remember one to Weardale and being amazed by how the older boys could remember the birds' names. I wanted to do that, to know the names. Another time we came up here to Aberdeen and stayed in the Youth Hostel. I was sharing binoculars then, and didn't have my own until I'd been birdwatching for three or four years.

'I left school at sixteen and went to work in the same crane-building factory as my father had done. I took an engineering apprenticeship and worked as a plater. A job for me then was a way to earn a living, to fund my hobby. By that time birding had taken over my life. Because of the job I was able to afford a car and was travelling to see birds and had started going abroad. Aged about twenty-one, I became interested in wider environmental and ecological issues and realised that I didn't want to work in a factory forever. So I went as a mature student to university.

'To begin with I had no more fascination with gulls than any other birds, though one of my regular places to watch was a gull site – the small fishing harbour in Sunderland. Then there were two events that got me focussed on gulls. At school, Brian Bates showed us pictures of a twitch he'd been on to Blackpill in south Wales to see a ring-billed gull. The concept of a rarity was new to me. I still remember Brian saying how difficult the bird was to identify: you have to look for their squinty eye, he said. Such an apparently mythical bird got me out trying to find one at the harbour, and so into rare birds and twitching more generally. The second event was related to the fact that I had by then started going to Scilly. One autumn, I met Peter Grant in the Porthcressa Inn signing copies of his new gull identification book. It was published in 1982 and had a ring-billed gull in it. I was hooked. I'd met a man who'd written a book and it was about gulls. It's still here in my office.'

Chris pulled it down from the shelf above.

'One way or another since then it's been gulls. In Aberdeenshire I do most of my gulling on my own. It would drive other people crazy, sitting in a car at Peterhead watching the quays all day.

'When Martin Garner's first papers on yellow-legged and Caspian gulls came out in *British Birds* in the late 1990s they really grabbed me; he had puzzled through their identification long before they were split taxonomically. I remember the feeling of shock: these forms are occurring in Britain but we've all been missing them. If you don't know something exists how can you look for it, how do you ever see it? Maybe I'd already seen lots, maybe I hadn't. I still remember that feeling.

'For Caspian, I work from herring gulls. They are my baseline. Caspian gulls have a strange quality: they're more aggressive than herring gulls but also I think more beautiful. Beauty is always a tricky concept to use for large gulls – perhaps elegant is the word. Then again they show a lot of variation between the sexes: sometimes it is hard to imagine they are the same species, as females are like large common gulls, while males can be huge brutish beasts. In mixed gull colonies in Poland and Germany they are increasing, and they are aggressive. But Caspian aren't common here, there's only been ten or so accepted records in Scotland, and I've only had one in Peterhead.'

I asked Chris whether he thought birders felt threatened by taxonomists splitting species that might be inseparable in the field?

'It's a big issue, and there has been much debate since the use of DNA in taxonomic decisions. Like most birders, I don't have the training to really query the genetic work, and I tend to trust the decisions of scientists who are the specialists. But then it turns out – and this causes confusion and frustration among birders – that there is still a lot of room for different interpretations of the genetics.

'There are still fascinating bits of the jigsaw missing. Often these are related to taxa that occur in remote areas. The geographic variability of birds has interested me for a long time. I am curious irrespective of whether they are separate species or not. With more birders travelling to unwatched or underwatched places we are just beginning to be able to look at some of these taxa and understand this variation.

'I've been going to east Asia in the winter to look at Vega gulls in Japan and Korea. There are also *mongolicus* gulls (believed by some to be a subspecies of Caspian gulls) in the winter. But not knowing where all these birds come from – where they breed – creates problems. I had a chance to go to Mongolia in the summer, with Andreas Buchheim, a German ringer who for years has been ringing and wing-tagging Mongolian gulls on their lake breeding grounds. His is fantastic work because we are now picking up where these birds go in the winter: Japan and Korea and China.

'By the time I went to Mongolia, I'd become interested in sound recording and using sonograms as a means to identify gulls. We've known for some time with Caspian gulls that calls can be diagnostic, so I was keen to try to record *mongolicus* long calls to see if they differ. In Malling Olsen's book *mongolicus* is regarded as a subspecies of Caspian; Martin Collinson's paper in *British Birds* on large gull taxonomy argued that it is not a Caspian but they were not sure where it sat. For convenience they left it in a Siberian clade, a group that also consists of Vega gull and American herring gull. I went to Mongolia with the view that Mongolians don't look like either of these, but let's see if I can record them and see how they sound.

'This takes me back to Caspians. For me, it seems that we're going through a full cycle. In the past we knew very little about them, then suddenly we were very aware of them. Now the population is growing, and the range is expanding. And its hybridising left, right, and centre – we might not see or know pure birds for very much longer in Europe. So we've had a window when we might have known what a real or pure Caspian gull looked like, but now this window is closing.

'It is fascinating but it's too big a puzzle: you're confronted with hybrids between Caspian gull and other taxa, when in fact we are not yet really sure how to identify pure individuals of these other taxa. In the UK, I am usually happy to call things at a species level. But there are lots of individuals for which I don't know what subspecies they might be. The more I look at gulls, the more I see birds I'm not sure about.

'There's a small dump at Batumi on the Black Sea coast of Georgia. It's a crappy place with good opportunities to get really close to gulls, for photos and sound recording. It's smelly and noisy. The guys who work there have an extraordinarily tough life: they spend their days raking through the rubbish for salvageable metal and plastic. I was on my own there on a couple of trips – totally odd to them, a figure of fun in many ways, but they were good people. On my last day on one of the trips I took a bottle of whisky onto the rubbish and we all shared it.

'It's very much the eastern edge of the range of yellow-legged gulls there in Georgia. They start to look more and more like Caspian gulls. As well as them, I saw quite a few Armenian gulls, lots of *fuscus* lesser black-backed, some *heuglini* (a full species for some), and possibly *barabensis* birds (regarded as another Caspian subspecies). Absolutely confusing and perplexing – I didn't know where anything had come from. That's when it dawned on me, at the Batumi dump, that in many cases you're not quite sure what it is until you hear a bird call. I realised that plumage can only be a small part of the story. I needed to hear them. When feeding they are most vocal, and tuning in to their calls is really important. So I'd spend all day at the dump, eight to ten hours, walking around with individual birds and waiting for them to call. And then, once I was sure, I'd take photos. This change meant that sometimes I had photos of just ten or fifteen birds a day, but at least I knew what they were.

'The Caspian gull thing – merging taxa – seems like it might also be happening to other groups of gulls. The Kumlien's gull, thought by some to be a subspecies of Iceland gull, is spreading from the east coast of North America and potentially invading the core range of the Iceland. It's doing well and seems to be occurring now on Greenland, which has always been the home of the Iceland gull. Pure Iceland gulls have long been known to winter in Britain, but now we're seeing lots of things that look like Kumlien's gull. We don't know whether there's been a major change in movement patterns of Kumlien's, and that more are now coming to Britain from Canada, or whether we're seeing birds that are breeding in mixed colonies in Greenland.

'Because of the relatively recent ice in the northern hemisphere, the gull stories here are not ancient, but they are dynamic and perhaps never more than now, readjusting to a changing climate. There's a lot of discussion about the ice-free Arctic seas and what birds will make of them. The Kamchatka subspecies of the common gull was found in Newfoundland a little while ago – perhaps that bird had come over the top.

'Funnily enough, all this thinking about what a species is goes all the way back to where I started, saying that I made a conscious decision not to work with birds professionally. Now, I often find that I am frustrated that I don't know their biology and the genetics deeply enough to understand certain things. But, you know, a species is a human construct. I've always been desperate to find a Baltic gull here not because it might be a different species but because I want to know whether I can identify one confidently in Scotland or not.'

Isn't that a beautiful, even a lovable, thought – this expression of such interest in an individual organism. One bird. There, that one, in the middle of that dishevelled gang of them, standing around on the quayside, like men outside a pub at closing time, I want to know *that one* not because it might be rare or different, no, *that one* because everything about it says who it is, and if only I could read all that is written into it I could know it as an individual.

★

Living most of my life in Bristol, I have spent almost fifty years looking at the muddy waters of the Severn Estuary and the Bristol Channel, but I have been out on them only twice. I once had an aborted trip to Lundy that turned into an unwanted cruise of the birdless upper reaches of the Severn. My second time on the water was a day trip to Flat Holm. A boat called the *Westward Ho* makes the crossing from Weston-Super-Mare a few times each summer. I always thought that given the choice I would elect to go to the neighbouring island, Steep Holm. Who wouldn't choose Steep over Flat? But Flat Holm has breeding lesser black-backed gulls – thousands of them, spending their seabird summer jammed up a tidal estuary.

I persuaded my children to come with me. We shipped with members of the Flat Holm Society, including Peter from Cornwall, a small man carrying a pilgrim staff. (I misheard his first introduction. Flat Earth, I thought he said.)

The Severn is itself flat in most weathers, but the mud that underwrites the water here is so present that it further suppresses the whole moving wetness. The sea is mud suspended. The land, you feel, is still in dispute with the water, and intends to carry on as land even as it is being carried away as silt. There is no salty tang as you cross the estuary. The spray thrown by the ferry's engine marked my notebook with dusty droplets that dried like a muddy emulsion.

On the island about two-and-a-half thousand pairs of lesser black-backed gulls were breeding. Their nesting area covers roughly half of Flat Holm. Between stands of nettles and bramble patches they had scratched out spaces and built shabby nests – scrapings of old vegetation, bowered with brittle seaweed and prized strands of assorted junk, plastics mostly. There was a cabinet in the visitor centre on the island that collected together some of the birds' prized dolls and cigarette lighters and bottle tops from previous breeding years. A condom was my best find.

We were some way into the season and the gulls' nests were now more teenage bedrooms than cradles for newborns. The chicks weren't especially keen on being in them and were wandering around and squeaking. They were dressed in spotted pyjamas of dirty-coloured down. Their feet and eyes and beaks were as clownishly big as they ever would be; all else was still catching up. If they strayed too far, the next-door neighbours shouted and sent them home. Sometimes the adults are less kind and kill the chicks. There were bodies here and there, squashed into the dirt, looking like sloughed off old slippers, ignored by all.

The adult gulls were good to see close up. The unlikeliness of family, of domesticity, seemed written into their expression, their awkward flights and high-stepping walks, their calls. They looked and behaved as if bewildered by their diligent attention to the obligations before them.

The red bill-spot was working. Lucian pointed out a chick pecking at its parent's bill and the bilious ooze forthcoming. He'd seen such spots before, though his friend who he was with thought the red was a ketchup stain from some pilfered chips.

There were hundreds of bones scattered all across the colony. I thought they were rabbit bones at first – I knew some gulls had developed rabbiting skills. But Dominic, a medical student and more familiar with the taphonomy of street food, identified them as chicken leg bones. The gulls carry back to Flat Holm the leftovers of human fast-food eats picked up in Weston-Super-Mare and Cardiff and elsewhere.

We walked to the shore that faces Wales. Adult off-duty gulls were perched incongruously in some elder trees; on the beach, new-feathered chicks were sitting on boulders looking at the water and beyond it to Wales. I wrote down the rings that I could read: F417 left leg blue, F418 left leg blue, F410 left leg blue. The birds didn't seem likely to be taking their data with them anytime soon. Something inside them all had delivered the whole nursery to the edge but what was to be asked of them next seemed too far-fetched. A calling to the sea? Why? And, in any case, what sea? They stretched their new wings and looked at the muddy slosh and, beyond that, to the darker grump of land and its drumsticks. In the time that we watched none did anything else but look.

We waited on another beach for the *Westward Ho*. The sky had clouded and its waning light muzzed up the horizons of England and Wales. Young starlings, dusty and matt, bundled about on a clifftop. Greenfinches wheezed in the brambles. The last lesser black-back I saw close up was a muddy-toned youngster, the colour of the Severn around the island. Something had gone wrong on a first or early flight and the bird had broken its wing and injured its neck. Perhaps it had crashed into the low cliff where I found it lying at the foot of a rust-streaked rock. It looked up at me with its dark eyes. The brown world was closing around it. My boys, kindly over-interpreting the moment, walked off to give me a minute with the moribund bird-child.

Basil

Viola Ross-Smith is the Science Communications Manager at the British Trust for Ornithology. She worked as a research ecologist for the Trust before that, studying seabirds, including gulls on urban rooftops and on offshore islands. Before that, she studied lesser black-backed gulls on Flat Holm, 'getting very dirty', working especially on the pecking responses of young birds when they were soliciting food from their parents. When we met at the BTO's offices in Norfolk she was seven months pregnant with her first child. She grew up 'without gulls' near Oxford and remembers properly noticing her first gull on a school excursion to Conwy in North Wales. 'I liked their noise.'

Breeding gulls are new in Thetford. When Viola arrived in 2010 there were none. That has changed. We went to an industrial estate at the edge of the town and, as we talked, herring and lesser black-backed gulls came and went above us non-stop. They were breeding on the roofs of metal sheds.

'The rooftop is like an offshore rock stack for them. There are no foxes up there. The town has a plentiful supply of food; it is surrounded by agricultural land and we're not far from the rich pickings of the pig fields near Livermere. Gulls are moving into urban spaces in Thetford as they have done elsewhere. As well as breeding herring and lesser black-backs these roofs in Thetford attract a winter roost and in the winter of 2016–2017 we had Caspian and yellow-legged and Iceland and glaucous gulls here. The connectivity of the gull world was visible too – we had ringed birds from Poland, Lithuania and Ukraine.

'During the twentieth century, gull numbers went up across the UK. They started breeding in coastal cities like Bristol, Cardiff, Aberdeen, Felixstowe and so on, and then they started spreading out. These rooftops represent a vacant niche. The lesser black-backs would have been migratory once, but that seems to be changing. They have less need to undertake long journeys for winter food. And these birds are natally philopatric which means many of them come back to breed where they hatched themselves, so if hatching success is high in urban areas, which it tends to be, and those chicks are well fed and not predated, they'll come back to where they started and the population will expand.

'They are still pioneers here, with just ten or twenty pairs, but they have begun and look set to carry on. In fifteen years time I think it will be a lot noisier here. That said, gull colonies can come and go with great speed. Lesser black-backed gulls started breeding on Orford Ness in Suffolk in the 1960s – their population was estimated to reach approximately 20,000 pairs in 1999 – but now it has dwindled to fewer than 400 pairs, and in some recent years not a single chick has fledged. There are predatory foxes there but we don't really understand how these colony buildups and crashes occur.

'There's still a lot to learn. I was part of a team that tagged some Cornish herring gulls in St Ives with GPS. One lived on a cinema roof, one on a supermarket, and two on local cafes.

We caught adult birds when they were sitting on eggs and fitted tags on their backs that sent details back of where they were. It was only a small sample but we still found differences in their foraging behaviour: none seemed to feed much in St Ives, not one was a chip stealer, one went far out into the Bristol Channel, and the others went inland to a farm to feed on invertebrates in the fields; the tracker showed one neatly going up and down a field following a tractor. If you'd culled the town's birds trying to stop the chip thief you may well have got the wrong bird. That bird could be living on a cliff like an old-fashioned 'proper seagull' but coming into the town to exploit its food resources.

'The more we know the more we are learning about the individual personalities of these – and all – birds. More and more characteristics that we think of as unique human qualities seem to be held in common with many other animals including gulls. It's humbling, I think.

'My colleagues and I also tagged twenty-five lesser black-backed gulls breeding at Orford Ness. Some of those left their Suffolk breeding site and flew to Farmoor Reservoir near Oxford and then on to Spain for the winter. They did this every year. The same route, the same stopover, and the same destination: it was a gull culture that the birds had apparently learned. It worked for them so they stuck to it.'

'My PhD was on lesser black-backed gulls breeding on Flat Holm – a marine or rural site – but these birds, the opposite of the St Ives herrings, were flying to urban areas like Cardiff and Weston-Super-Mare to feed. Numbers on Flat Holm were quite good when I was there – more than 4,000 pairs of lesser black-backs in 2009 – but in the space of a few years the population has almost halved with lesser black-backs down to 2,445 pairs in May 2017. It's not a good sign. They are going down the tubes. Landfills are closing. The Flat Holm birds regularly fed at the landfills in and around Gloucester, but they were deterred when a falconer was employed in 2011. Other sites are going the same way – gulls are either prevented from foraging there or the landfills themselves are closed and covered. Though they are island nesters and are surrounded by a sea of sorts, the Flat Holm gulls remain terrestrial foragers. The Severn is not calorifically rich. There are few alternative food sources without landfill.

'The birds on Flat Holm knew me. If someone else came into the colony they reacted in a different way. They remembered me from one year to the next. I named a few. Some were shy. Some were voracious. Some were flighty, some perched on my head. Basil was a slow dive-bomber; he even had a slow-motion call.

'What is interesting is our getting closer to being able to see the variation between individuals and the personalities in the birds, the way some are docile, some aggressive, some are in stonking good condition and others worn out, some are habitual cannibals. At a wider level, these variations show the width of the ecological niche the species can occupy.

'The idea of species isn't that important to me, I am more interested in ecology and behaviour; all the species names changed as I was writing up my PhD but I don't really care if the Baltic gull is a full species or not, nor is my interest to do with the length of the mirror on the second primary . . . Things are happening to the gulls' environment and the birds are adapting to these human changes; *this* is interesting. Gulls are dynamic birds and fast adapting. They show a huge range of responses – behavioural plasticity – to the world they are living in. A pair of breeding lesser black-backs might split up in the winter with the female going to Portugal while the male remains in Norfolk. What's going on? A Dutch study has shown four different migration strategies in lesser black-backs. There are all sorts of things coming up. It turns out that birds that stay near their breeding sites might actually cover the same distance across a winter as birds that travel far away; the stayers do more wandering, the ones that go to Mauretania are sitting on the beach. We don't yet know why they choose one of these strategies. Young birds go further in their first couple of winters, like backpacking teenagers, and then they fix on whatever strategy they've adopted and stick to it. But did they learn this or is it innate? Are they communicating as living individuals or are they genetically driven? We don't know yet.

'It is odd isn't it: a lot of people love the sound of gulls at the seaside but in the cities they hear the same sounds as raw and aggressive. I'm not sure about all this overhyped scare-mongering. It's hard to believe how many times I've been asked whether gulls are getting bigger year by year (they aren't!). It is us making the story, not the gulls. All the chip-taking and ice-cream theft is a feeding opportunity, not an attack, that's all. If we made it more difficult for them to get into rubbish bins they'd go elsewhere.

'Gulls will protect their young, and fledged chicks on the ground often blunder into people, and their parents will bomb humans, but they strike with their feet, not peck with their beaks. Gulls don't do revenge attacks. Nor do the birds provoke the antipathy of, for example, Dutch people or French people in the way that they do here. It seems a particularly British thing. And some of these reported attacks seem a bit far-fetched to me. Some gulls do specialise in taking mammal prey – rabbits, rats and moles – but they are taken themselves by foxes (I've had study birds decapitated by foxes) so I think most pet dogs would be a threat to them rather than a meal. And children? Well, I'm not worried for my baby.

'It's also important to remember that we're responsible for all this. We've thrown so much edible stuff away; we've driven the gulls from their traditional breeding sites; we've discarded fish guts, providing gulls with a ready, predictable food supply and then we stopped, causing local breeding colonies to collapse as gulls that could not find food elsewhere were unable to provision their chicks. The same now is happening with landfill.

'Gulls are caught in an ecological trap with us. And what will happen next depends on how we manage our urban environments. If we continue to leave rubbish around, the birds will feed in these environments. The cause and effect relationship is well established between the appearance of landfills and a boom in gull populations; so, it would be odd if the phasing out of landfill food waste didn't have the opposite effect. Away from food waste sources it is hard to know how they'll manage. There might not be seagulls on sea cliffs any more. The herring gull is on the red list of conservation concern; other gulls are on the amber list. But these are based on rural areas, and they are doing better in urban areas. Maybe our cities will save the gulls.'

Age of Iron

Many of the kelp gulls that feed at the Coastal Park Landfill in Cape Town in South Africa breed next door in the beach dunes of False Bay. The sea there often runs brutishly strong, blown furiously by south-east winds that arm themselves far away in the Southern Ocean to mount harsh campaigns all along the South African coast. On many days, pelagic birds like albatrosses, prions and giant petrels are pushed towards the shore. Surfers enjoy the False Bay waves and great white sharks thrive in the swell and swing of the broad jaw of sea, where they eat the fur seals that pup on the offshore islands. The wind hits hard into the dunes, sending sand in prickling slaloms across the road, and forcing all the incubating or brooding gulls to turn and face the blast so that they or their chicks don't get blown away.

It is heavy weather and a hard living, and yet this place is very much part of the city of Cape Town. The gulls lift off from staring out storms and they can wash in the pools of the city's sewage works, just inland from the dunes, or they feed next door at the dump. The Cape Flats townships are gull-minutes away, as are the fish-and-chip cafes of Muizenberg. I've watched kelp gulls hang around the blood oils of a shark-killed seal in the bay, and I've also seen them take smoked *snoek* and *slap chips* from people eating on the front.

My wife is South African and we live in Cape Town for part of each year. The gulls here are simple and few; southern Africa is a long way from any *larid* hotspot. Freshwater places attract grey-headed gulls, and the marine edges have Hartlaub's and kelp gulls – one graceful, the other a big bruiser. I've watched kelp gulls dining with great white pelicans on the city dump, one pelican swinging a bursting supermarket carrier up and into the bin of its lower beak, two kelp gulls dabbing at its dribbles. And I've counted the Hartlaub's gulls on my local beach at Scarborough, as they made me feel homesick for the black-headed gulls of the north whose lives they seem to ghost.

There are a few other more occasional species to look out for. Sabine's gulls that have bred in the Canadian arctic cross the Atlantic to spend their winters off Cape Town's western shores. At Sea Point, a white suburb of cream-painted apartments towering like Miss Havisham's bride-cake, we stopped on the beach and peered through a telescope. A large flock of the fluttering small gulls was working over a hidden sewage upwelling. They hovered like terns, and only when they shifted side on and began to move further out into the bay past Robben Island could I see the magic geometry of their wings: a white triangle, a black triangle, a brown triangle.

I tried to imagine what a passing Sabine's gull might have meant to Nelson Mandela during his offshore incarceration. The people of South Africa – whitish, blackish and brownish – are a lot more complicated than its gulls. Divisions and systematics, attempts to box and label the uncategorisable, bogus racial science, and the pernicious policing of every human type – all is deeply written into the country's history and is not yet beaten out.

You might attempt a taxonomy of this town. It is a highly segmented city. The architecture of apartheid still prevails, and separation remains the watchword for too much of life. Mention any part of Cape Town to any citizen and they will know the colour of that district. In the daytime, the business of the city means that it mixes, as gulls might fly from a morning above the wine gardens of Constantia to an afternoon over the dust yards of the Cape Flats; in the night the city sleeps largely with neighbours of the same skin tones.

J. M. Coetzee's novel, *Age of Iron*, is a book to hold against Sea Point and Robben Island. A visionary parable, it is a dark projection of Cape Town life after apartheid-era South Africa. It was written as that world was dying, but imagines its apocalypse. The last page of the novel notes that it was composed between 1986 and 1989. The book is much concerned with *rubbish* and anything that might be conceived as that.

Elizabeth Curren, an elderly teacher of classics in Cape Town, finds a rough sleeper in an alley down the side of her garage. The same day she receives a terminal diagnosis from her doctor. The rough sleeper, Vercueil, has moved into a 'dead place, waste, without use, where windblown leaves pile up and rot'. He's a derelict and a drinker, but Elizabeth Curren's life, as she retells it through the novel, turns out to be far more wrecked than the alley and its lodger.

Her house is full, she says, of 'rubbish and dead memories'. In a letter addressed to her daughter in America, she describes how she finds the nearby natural world beautiful – 'these seas, these mountains' – but she cannot love its people, least of all herself. Her body is dying faster than her mind. And she hates it:

> I look at my hand and see only a tool, a hook, a thing
> for gripping other things. And these legs, these clumsy,
> ugly stilts: why should I have to carry them with me
> everywhere? Why should I take them to bed with me
> night after night and pack them in under the sheets,
> and pack the arms in too, higher up nearer the face,
> and lie there sleepless amid the clutter? The abdomen
> too, with its dead gurglings, and the heart beating,
> beating: why? What have they to do with me?

Nor can she care for anything of the nearby world that people have made. There are rabbit skeletons and rotting fruit in her back garden. She gives Vercueil a job as a handyman and fixer-up, or tries to:

> 'Anyhow, do what you can to bring it back under control,'
> I said. 'So that it doesn't become a complete wilderness.'
> 'Why?' he said.
> 'Because that is how I am,' I said. 'Because I don't
> mean to leave a mess behind.'

But her tribe already has. Coetzee writes brilliantly of the white occupation of South Africa, the land grab of European settlers and their attempts to declare ownership of the very earth, and to seed, with themselves, its fruitful soil.

We know Vercueil is grimy but we don't know what sort of man he is. His name sounds Afrikaans and that might suggest he's a Coloured (the South African term, still in use, for people of mixed race). His name also sounds like Virgil. And, as Virgil guided Dante around hell, Vercueil becomes a guide to Elizabeth around the chaos of the failing city. Might he offer her a way out?

Florence, Elizabeth's black domestic servant, is horrified by Vercueil and challenges her boss.

'He lives here,' said Florence, 'but he is rubbish. He is good for nothing.'

That *nothing* might be the crux of the book: what it means to be not yet or no longer, to be nothing, rubbish, annihilated, nonhuman, useless, unwanted. It is also the making of the end of Elizabeth. To be regarded as nothing is a cruel fate and yet, the way we have worked the world, wrecked and ruined its people and its life, makes nothingness potent too. 'Beneath it all, the desire for oblivion runs,' says Philip Larkin in his poem 'Wants', and *Age of Iron* trades Vercueil's worthless nothing for Elizabeth's species-shame and her own sought extinction. The world needs purging. Some burning, some smelting must come. Elizabeth describes how cremation is to be preferred to burial – 'to burn and be gone, to be rid of, to leave the world clean' – and the same applies to the terminal politics of the white state, 'this country too: time for fire, time for an end, time for what grows out of ash to grow.'

'*Jou moer!*' Vercueil replies to Florence. In Afrikaans, *moer* is a deep and tangled word with a complex linguistic DNA. *Marsh*, *mud*, and *mother* all lie somewhere within it; and at its heart is a paradox – it simultaneously signifies the beginning of life and its end, the seed and the shit. It is like *dust* in this way. Literally it means 'your uterus' and as such it's a way of saying 'you cunt' whilst hanging onto the original gynaecological or procreative charge of the noun, even as it is used as a hostile expletive. A double arming. One online dictionary gives other rhizomatic meanings too: 'dam, draff, dregs, faeces, a female grooved screw, foot, grounds, lees, matrix, mother, nut, seed potato, settlings, and womb.'

Elizabeth defends Vercueil.

'He is not a rubbish person,' I said . . .
'There are no rubbish people. We are all people together.'

Florence describes her own children as the opposite of rubbish; they are 'like iron' even as they are persecuted and hunted down by the government. Later Elizabeth records how she, as a white person in South Africa, feels like she is walking on black faces: 'Millions of figures of pig iron floating under the skin of the earth. The age of iron waiting to return.' This is a book, then, in part about *rust*, about what happens when iron meets the rain coming off that southern ocean in a South African winter, when the novel is set, and what happens when that iron met the stronger metal weaponry of the country's former rulers.

Vercueil takes Elizabeth to Gugulethu, a Cape Town black township. It is still there today. It was the home, in the apartheid mind, to the country's human rubbish. The visitors see the torching of houses and a dead body, freshly killed by the police. It is fifteen-year-old Bheki, the son of Florence. His iron has failed him.

Surrounded by a hostile crowd, Elizabeth is forced to speak. She is probably the only white woman for miles, the only professor of classics for even further.

'These are terrible sights,' I repeated, faltering. 'They are to be condemned. But I cannot denounce them in other people's words. I must find my own words, from myself. Otherwise it is not the truth. That is all I can say now.'

'This woman talks shit,' said a man in the crowd. He looked around. 'Shit,' he said ...

'But what do you expect,' I went on. 'To speak of this' – I waved a hand over the bush, the smoke, the filth littering the path – 'you would need the tongue of a god.'

Elizabeth is not that, and things get worse. Hell comes closer to town. Everything is falling apart. She ends up sleeping rough, like her Virgil, out in the city, as many there have, and still do. Marauding children attack her. Vercueil picks her up. Coetzee writes as if he'd just been out for a day to Coastal Park:

> With his high shoulder blades and his chest narrow as a gull's, I would not have guessed that Vercueil could be so strong ... Will this be how the story ends: with being carried in strong arms across the sands, through the shallows, past the breakers, into the darker depths? ... When would the time come when the jacket fell away and great wings sprouted from his shoulders?

This is not *Jonathan Livingston Seagull*.

In one of Elizabeth's last exchanges with Vercueil she wonders if the gull might indeed take her to the dump by the beach.

> 'You should sell these things if you don't want them.'
> 'Sell them if you like. Sell me too.'
> 'For what?'

'For bones. For hair. Sell my teeth too. Unless you think I am worth nothing. It's a pity we don't have one of those carts that children used to wheel the Guy around in. You could wheel me down the Avenue [towards the Parliament building] with a letter pinned to my front. Then you could set fire to me. Or you could take me to some more obscure place, the rubbish dump for instance, and dispose of me there.'

*

Fire is a powerful ecological driver in the Cape landscape. Much *fynbos* vegetation, unique to the area, only flowers after a burn. It is required for life. In our village on the Cape Peninsula, volunteer fire-wardening is taken very seriously. The hills can flame periodically – they need it – but we hope our houses won't burn with them. On the day I finished reading *Age of Iron* another strong south-east wind pushed a hill-fire close to us; Claire was called to assist at the fire break. I stayed home and watched the smoke greying the sky and rubbing out the sun, and soon ashy smuts started to rain on my white pages.

Split ·

Tring is where the Natural History Museum keeps its scientific collection of birds. My visit in 2017 was my first for thirty years. My last had been in late April 1986, on the day when the news of the Chernobyl explosion reached Britain. In 2017, the year of my return, the museum was wrapped in scaffolding and plastic; its 1960s cladding was coming loose and needed repair. Even mausoleums have to be kept up. Inside, the museum is almost unchanged: skeletons and birds preserved in alcohol (the 'spirit collection') are on the ground floor, smaller birds (passerines and hummingbirds, kingfishers and trogons) on the first floor, somewhat larger birds (non-passerines) on the second, and big ones on the third. (Eggs are stored in another building, which also houses giants like kori bustards, pelicans and albatrosses.)

Security is tighter than it was. First-time visitors must show their passport and a gas or electricity bill, and you must shift the contents of your rucksack to a see-through plastic bag. Not long ago a thief cased the joint on a public visit and then broke in, putting a brick through a window, to steal some quetzals and birds of paradise. He plucked the birds' reflective feathers and, when caught, was busy selling them online for high prices as specialist flies for fishermen.

Such unwanted recycling of the museum's great keep of feathers is an unfortunate theme at Tring. Most of the collection is made up of *skins* – the gutted and preserved bodies of 750,000 birds. These are not like front-of-house for-the-public specimens, cased and mounted in lifelike postures on a mossy branch or a rocky perch with glass eyes and repainted feet, legs and beaks. Skins are back-of-the-shop reference materials for study. They lie, awaiting investigation, in the dark of a drawer, stiff and straight, their wings folded tight to their bodies, a dab of cotton wool doing for eyes, their beaks often threaded shut, a label tagging them tied to their feet, and sometimes a stick jammed up their insides.

In this form, should you desire them, they could readily be slipped into a pocket and pinched. It is cold in the collection – cooler temperatures preserve the bodies and keep the moths at bay, and in the days of overcoats – Ulsters, Invernesses, Dusters – entire trays of smaller birds, or one or two bigger ones, every holding of a rarity, could pad out a gentleman visitor with undetected down.

One famous thief of the early twentieth century still clouds the collection at Tring: Richard Meinertzhagen's stealing was bad, still worse was the way he gave back what he had lifted. He collected birds himself around the world as a sideline to a life as soldier, adventurer and spy. But he also took a magpie shine to others' birds. In the museum, he pocketed specimens from the collection, rewrote their labels with fake or enhanced details, including his name, and bequeathed them to the place from which he had stolen them. Only recently has the scale of his robberies and his recycling been fully appreciated. The toll is still being taken. The missing birds – a space on the tray where the specimen should be, according to the accessions inventory – have been mugged and dressed up as another. Minds fog trying to tidy the mess.

Meinertzhagen's ornithological deceptions might describe his own character. He told lies about much of his life – took a bit of owl from here, a bit of bustard from there, and put his name to the rehash. The litter from his junking of the truth has blown widely through Tring's collection; its trail of chaotic disinformation has undermined what was thought to be known about a number of species, and fractured many an account that the specimens in the drawers have told. His wrecking has not only confused understanding of how the dead once lived, but also how some of their living relatives might survive. His jiggery-pokery with skins of rare species meant birds were subsequently searched for in the wrong places and were missed, or even presumed to be extinct.

I didn't go to the museum to look for Meinertzhagen: he pushed his way through the ruins. His own remains are elsewhere, but he's still disturbing the dead. The ornithologist Pamela Rasmussen and Robert Prŷs-Jones, the retired Head of the Bird Group at Tring, are at work on a comprehensive Meinertzhagen fraud dossier (Brian Garfield and Mark Cocker have told the detective story to date). Meanwhile, tray after tray of skins at the museum that I pulled into the light had handwritten notes added to them warning researchers that among the birds before them were 'Meinertzhagen specimens', innocents themselves but toxic to science unless handled with great care.

⋆

I drove to Tring with Nigel Collar. As well as being a research conservationist for BirdLife International, where he has written the bird *Red Data Books,* doomsday directories for threatened species, Nigel is one of the authors of a new checklist of the 10,000 species of birds of the world. Its two heavy volumes weigh down my desk.

Thirty years ago, I was working as Nigel's assistant on the first installation of the Red Data series, *Threatened Birds of Africa and Related Islands.* I spent three years in Cambridge cycling between various libraries and a photocopier, where I scanned Victorian and early twentieth-century journals of natural history and exploration. Nigel wanted everything that anyone had ever known about 200 or so species of African birds.

To prevent the Madagascar serpent eagle becoming extinct, the eastern rainforests of the island must be preserved forever. That could not be mandated from a caravan doubling as an office on the outskirts of Cambridge. But if we could gather everything anyone had ever written about the serpent eagle, we would surely be better placed to help secure its future.

Nigel knew the authors of contemporary studies and was promoting new searches and expeditions, but I was his retriever for the old stuff. The dusty stuff. Ornithological notes from a journey of 'twelve hundred miles in a palanquin' described in the *Antananarivo Annual* of 1892. Missionaries' diaries and wormy letters. Uncatalogued acquisitions as from the papyrus dump at Oxyrhynchus. The eccentricities of a few hundred Europeans putting pen to paper about what they had seen of birds across the continent, from bald ibises in Morocco to jackass penguins at the South African Cape.

I was the first to open many of these books and journals. Older volumes were often uncut. There, the never-read described the barely seen. A record of a sighting of a single forest robin or a rock thrush would spring from beneath my paper knife. How many people had thought of that bird between the day its watcher had sent his material to Europe and the day I cut its printed pages?

★

I had come with Nigel to Tring this time because I wanted to watch him working with the dead, and to ask him about his new checklist.

'I guess I must have split more bird species than anyone living,' he said as he pulled the first tray of the day from the shelves of a tall white cupboard and carried it, like a baker with hot bread from an oven, to a table by a window. 'But that's only because I've looked at more cases. I still reckon I'm as much a lumper as a splitter. I just want to see the evidence.'

The overhead lights in the museum only come on when something moves beneath them. The dead on their shelves are going nowhere and lie down in darkness. But as you walk between the specimen cupboards, along the narrow passageways, strip-lights crackle brightly above you.

As Nigel gathered more trays I looked at the gulls. They were mostly white but many appeared a little *off* against the white cupboards in the white trays under the clinical lighting. The drawer of ivory gulls looked like an aerial view of a pod of belugas.

In one cupboard there was a handwritten note taped to the inside of the door questioning some of the attributions and the organising of the skins within. Sometimes when a new configuration is agreed a new label is added, but the original label remains. Knowledge accumulates but is not overwritten. The trays are marked in black pen with the birds' subspecific names and their origins. These are revised to indicate current thinking, and many are smudged. It is still a busy time with the gulls. I found a note from Martin Garner, the amateur but expert birdman who was interested in field identification, especially of complicated families of birds. He was among the finders of the first Caspian gull at Mucking dump in Essex in 1995 and was still working on gulls with Chris Gibbins and others when he died in 2016. He'd opened this cabinet, looked at its dead and now he was too. Thus do the once-living send messages out of the dark. I had to shut the door.

I walked down a corridor and found a resurrectionist at work. Nigel was peering at a tray of laughingthrushes, assessing whether a subspecies found in Sumatra could merit full species status. The saving of the world might be at stake.

Nigel talked as he examined the grey and brown birds laid out before him. The laughingthrush is thrush-sized and looks like one, but it's a babbler with a fantastic voice – making it much sought after for bird-singing competitions in Indonesia.

'This is *Garrulax palliatus*. It's only found in Sumatra and Borneo. There are two subspecies; the one in Sumatra has been hit really badly by the bird trade: they are birds of the understorey and they travel in big groups, so when the trappers go in with tape recordings and nets a whole flock can be caught very easily. A mountain's worth of birds can be wiped out in a matter of days.

'The nominate form *Garrulax palliatus palliatus* has become extremely rare, so one question is: how different is it from the form *schistochlamys* that occurs in Borneo? How distinct are the birds in the two populations and should we treat them as separate species? You can see, when they are side by side, as here, that the grey on the crown and mantle of the Borneo form goes further down its back. After this quick squiz I think we could be looking at two species, but we are only comparing a few examples.'

Nigel asked me if I thought the bills of the two laughingthrush subspecies looked different. Put on the spot, I wasn't sure. Perhaps.

'There are cases where forms of birds have diverged between Borneo and Sumatra and the different populations have come to be regarded as separate species, but there are also cases where the differences aren't that great – I'll measure the bills in a minute – and the birds are regarded as subspecies or even without subspecific difference at all. Borneo was once connected to Sumatra when sea levels were lower, but the isolation of the two islands over time has led to divergences in their once shared birdlife.'

So it's a question of time?

'Time is a factor, but it's not the only one. In some cases evolution can take place very rapidly. Sometimes you see a several million year separation in bird populations and they are genetically very distinct, but when you look at them there are hardly any differences at all to see. Are they species or not? My view is that if *they* can't tell themselves apart, then the birds are still probably reproductively compatible even after several million years, and therefore not separate species. In other cases there is really rapid divergence, even after a few tens of thousands of years. It is very contextual and very fluid.'

Nigel remembered the common name of the bird we were looking at. Up until now the Latin was all we had.

'It's a Sunda laughingthrush!'

There was a Meinertzhagen specimen among the birds in the tray. Not too bad a crime this one – the specimen label misidentifies the bird, but he'd left the original collector's label and hadn't tried to claim that he'd shot it himself.

Another laughingthrush had a label saying it was collected on a 1932 Oxford University expedition. That would mean, Nigel said, that Tom Harrisson, the man who moved his collecting and categorising interests from birds to people and went on to found Mass Observation, might well have held the bird.

Nigel checked his laughingthrush findings and his thoughts with the entry that he wrote himself in 2007 on the Sunda bird in the *Handbook of the Birds of the World*. 'Maybe,' he said.

*

The big noise in taxonomy these days, is genetics.

'Molecular analysis has shown some fantastic things about the relationships of birds at the higher levels. It has reshuffled the deck completely. In all the old handbooks and field guides, pigeons are next to parrots and parrots next to cuckoos, with the implication that there is some connection between them. But we now know that pigeons are totally on their own, separate from everything else, with no close relatives at all, whilst parrots turn out to be very closely related to passerines, and cuckoos turn out to be related to all sorts of birds we would never had imagined. On the newly determined phylogenetic tree, cuckoos are close to bustards. This seems impossible to believe: cuckoos are mostly parasites, they have two toes pointing forward and two back, bustards are large birds of open country with three toes all pointing forwards. No one would have conceived of a connection.

'But when genetics investigates the status of subspecies it gets much more difficult to be confident about its findings.

Lots of geneticists have declared a measure of two or three per cent molecular difference between one taxon and another as enough to split them into species. But this principle is, I think, unsafe because cases are emerging where even large levels of genetic distinctiveness between populations of birds that look very like each other do not reflect the existence of two species. Common redstarts are an example: there is a line down Europe where the difference in genetic composition on either side is, for whatever reason, something over five per cent, and yet where the birds meet they are totally reproductively compatible. The same happens with ravens in North America – the birds on the Midwest plains and the birds in the Rocky Mountains are just under four and a half per cent genetically different and some raven people once argued that that made them different species, but when the two populations were studied where they meet they were found to interbreed perfectly, with their offspring fully fit.

'Even in something as seemingly objective as genetics it seems that taxonomic decisions can come down to individual judgement, and I'm not yet convinced that genetic evidence can be trusted to deliver an incontrovertible decision on species. I don't think it's a silver bullet – yet.'

We went back to the collection, this time to the hummingbird quarter, and the part of a tray where all the museum's holdings of the bearded helmetcrest are kept. Each was like a tiny caddy or finger-puppet, beetle-bright and made of golden, green and bronzy threads, and around each bird's needle-bill was a mini-beard and a mini-crest of various iridescences. Listening to Nigel, and watching him turn in his fingers the tiny birds so they caught the light, all their deadness disappeared for a moment.

'*Oxypogon guerinii*: a species made up of four subspecies. It's found in a few small areas of Colombia and Venezuela. There's one form with a green beard, one with a sapphire-blue beard, there's a white bearded form, and one that we don't have here has a multi-coloured beard. When I first looked at this tray, the blue-bearded form, *cyanolaemus*, from an isolated volcano in northern Colombia, hadn't been seen since the month of my conception in February *1946*! Our recognition of it as a full species sent people looking for it in the páramo vegetation of the volcano. They found a tiny surviving area – most of the native vegetation was gone, burned to encourage grass for cattle grazing – and three birds. There is now a chance of saving this bird because of this revision to its taxonomy. If we hadn't split it, no one would have looked for it, or cared about it. Now with luck it has a future.'

Why do we give more attention to species than subspecies? Would you split a species in order to save it?

'No.'

But do you love a species more than a sub?

'I don't want anything to go extinct, but conservation inevitably follows the taxonomic hierarchy. Species are by definition more distinctive than subspecies, and I want to be sure that conservationists target the most distinctive things. If an art gallery was on fire, would you rescue all twenty-five of its Andy Warhols ahead of trying to grab a Rothko and a Rauschenberg?'

Outside, through the window, beneath the yews and firs at the back of the museum, I could see the only living birds I'd seen all day: two shadowy male blackbirds. I couldn't hear them, but they would soon be making the *chinks* and *chooks* of their going-to-bed sounds. Nigel hurried, at my request, to get some blackbird trays down. We peered at rows of regular specimens and a tray of aberrant birds, albinos and other oddities.

'I usually don't look at common birds when I am here; it brings you up short to realise what they look like as specimens. With these ordinary blackbirds, though, already the most obvious thing is the variability of the females.'

We fell to talking about the unheard birds, and the spring evening just a few days hence when we could come down our respective streets to our homes and be serenaded by a dark scarf of blackbird song slung from one territory to the next: 'a piece of utter magic', Nigel said, 'that makes life so much more worth living.'

holding air in descent

Funes the Memorious

Michel Foucault began *The Order of Things*, his excavation of the archaeology of our thought by quoting Borges. The blind Argentinian librarian-writer was himself quoting an entry (presumably invented by him) from 'a certain Chinese encyclopaedia' that offered a fantastic animal taxonomy. It is funny and well known, but it is worth quoting again. All animals are to be divided as:

(a) belonging to the Emperor, (b) embalmed, (c) tame,
(d) sucking pigs, (e) sirens, (f) fabulous, (g) stray dogs,
(h) included in the present classification, (i) frenzied,
(j) innumerable, (k) drawn with a fine camelhair brush,
(l) etc, (m) having just broken the water pitcher, (n)
that from a long way off look like flies.

Borges's list made Foucault laugh, he said, but it also made him *see* – it was a laughter that shook him at the very 'impossibility of thinking *that*'. The fictitious Chinese encyclopaedist thought like that, and we think like this and what, if anything, is the difference? The world *is*, and then the world is as we *say* it is.

Borges's writing makes great play with this fundamental classificatory distinction: the telling of the world might be all that we can say of the world but it is not identical to it. His interest in this puts him among the great literary taxonomists. Deluded inventories and cracked systematics are everywhere in his prose. The idea of a blind librarian – an organiser who cannot see what he is organising – further underscores the hybrid or feral status of much of his work. Who is the *I* in it? Did he make it up? There are inventions, borrowings, thefts, maybe some autobiography.

Borges compiled his own encyclopaedia of animals: *The Book of Imaginary Beings*. And in his collection *Labyrinths*, a mix of fictions, parables and essays, we find 'Funes the Memorious', a story about the ways in which splitting the taxa of the world ever more precisely could be bad for your health. I am keen on this story in part because my introduction to Borges came from the only taxonomist I know well: Nigel Collar gave me *Labyrinths* for my twenty-fifth birthday.

Ireneo Funes was from Fray Bentos in Argentina. As a teenager he was paralysed after being thrown from a horse. Before that, he'd been, he says, 'blind, deaf, addle-brained, absent-minded'. But he was injured into an extraordinary condition, a vertiginous gift, such that, when he came around after his accident, his perception and memory were 'infallible'. Now he lies in a dark room in his mother's house and remembers everything. This is James E. Irby's translation:

> We, at one glance, can perceive three glasses on the
> table; Funes, all the leaves and tendrils and fruit that
> make up a grapevine. He knew by heart the forms of
> the southern clouds at dawn on 30 April 1882, and could
> compare them in his memory with the mottled streaks
> on a book in Spanish binding he had only seen once and

with the outlines of the foam raised by an oar in the Rio Negro the night before the Quebracho uprising. These memories were not simple ones; each visual image was linked to muscular sensations, thermal sensations, etc. He could reconstruct all his dreams, all his half-dreams. Two or three times he had reconstructed a whole day; he never hesitated, but each reconstruction had required a whole day. He told me: 'I alone have more memories than all mankind has probably had since the world has been the world.' And again: 'My dreams are like you people's waking hours.' And again, towards dawn: 'My memory, sir, is like a garbage heap.'

The enduring specificity of Funes's memory means nothing is lost to him, but also that he is trapped in his retelling of life and can no longer live himself. Time is shipwrecked around him. He is like Casaubon in *Middlemarch*, seeking 'the key to all mythologies' and not realising he was turning to dust as he did so. Funes's injury made him virtual long before the digital age. His memory is like a dump where everything has been thrown away but nothing has perished. Each instance becomes a unique record for him, but it is only a record. He is incapable of general thought or, what we might call *lumping*.

Not only was it difficult for him to comprehend that the generic symbol *dog* embraces so many unlike individuals of diverse size and form; it bothered him that the dog at three fourteen (seen from the side) should have the same name as the dog at three fifteen (seen from the front). His own face in the mirror, his own hands, surprised him every time he

saw them ... Funes could continuously discern the
tranquil advances of corruption, of decay, of fatigue.
He could note the progress of death, of dampness. He
was the solitary and lucid spectator of a multiform,
instantaneous and almost intolerably precise world.

Funes is the ultimate taxonomist, a demon splitter. In his
memorious mind, every individual is distinct and separate.
There need be no species, no subspecies. Each animal is itself
and itself only, joined to its moment but not to anything
else. But the world will be undone by such antic generosity.
Borges returned to this idea in a fragmentary tale called 'On
Exactitude in Science' which (as translated here by Andrew
Hurley) describes the cartographers of an empire seeking
perfect representation and drawing larger and more detailed
maps until they...

struck a map of the Empire whose size was that of the
Empire, and which coincided point for point with it ...
In the Deserts of the West still today there are Tattered
Ruins of that Map, inhabited by Animals and Beggars ...

You cannot work at a 1:1 scale or take a day to recall a day.
To be able to think means to generalise, select, categorise, to
suggest that individuals might be species and to be happy to
make families. Funes, on his own in the dark, honours every
moment he can recall, every dog of every afternoon, but he can
join none of them. He is too busy with his annotations. The
world teems. His is a mad replica. Like braille script beneath
blind fingers he can feel every memory. Each is textured,
raised up, by its own particularity. But such tactile recall is

disastrous. Remembering everything prevents you knowing the value of anything. Funes's miraculous skill is also a curse. He is a backup drive to the world he has known. Everything has been saved, but without any sense of why it was worth saving. He cannot see the wood for the trees.

> Locke in the seventeenth century postulated (and rejected) an impossible language in which each individual thing, each stone, each bird and each branch, would have its own name. Funes once projected an analogous language, but discarded it because it seemed too general to him, too ambiguous. In fact, Funes remembered not only every leaf of every tree of every wood, but also every one of the times he had perceived or imagined it.

*

I had a fever. My sleep was hectic and broken. One dream in particular could not be finished or purged. Though I woke fully out of it, back it came the moment I shut my eyes. And it snagged and juddered there, melting like a film stuck in the projector-gate in the cinema of my hot head. It was a gull dream. It said there were *pure gulls* to be found in a square of sea that was labelled 5/6 on a grid on a globe in my mind. This was far offshore. The sea there was terrifyingly rough and lent against me heavily and cruelly. But the gulls were good and came down to the seethe of the swell and walked on it, as petrels can, in a white flurry of strong feathers. I thought it was stupid that they could pass for angels. I didn't believe any of that, but there they were and, somehow, they were good for me.

There is a square of blue sea that we have sort of called 5/6. After my dream, I looked it up in an atlas and realised I had once been there when awake and well. In British waters the most marine square between the 5th and 6th lines of longitude west of Greenwich lies in the Atlantic where the ocean opens to the north-west of Cape Wrath. It is old seagull territory or fetch.

Having crossed that square by boat, we headed on to North Rona, the last rock before Rockall. On the island there were breeding storm and Leach's petrels that we tape-lured to come close to us, there were bonxies that grunted their heavy-keeled disapproval low over our heads when we trespassed their breeding lawns, there were fulmars tucked under every rock. Auks teemed on the cliffs, and killer whales made the seals whimper just offshore, but no one looked quite like they owned that place and the sea around it as much as Rona's great black-backed gulls. Having hatched, as it seemed, from the storm-grey waters offshore, they had planted themselves with proprietorial weight on the rocks that rimmed the island. From there, with slow turning heads and cold knowing eyes, they looked about them, past our stick figures bending in the wind and far out to sea where the truth lies.

Night Watch

The last tray I looked at in the museum at Tring was for me. I'd been documenting the day, taking photographs of the gulls and recording Nigel Collar's words. My tape machine was running but as we opened the drawer of Madagascar nightjars, we both fell silent before the rows of brown bodies. They might have been asleep. Thirty years before, I had taken the same white plastic tray from the dark cabinet to a sunlit table to read what was written on the labels attached to the nightjars' short stiff legs. From that information and more like it, I wrote a book.

As I worked with Nigel on the African *Red Data Book*, he suggested I should also collect as much literature as I could on the endemic birds of Madagascar. Some of these were going into his big book as threatened species, but the island has an extraordinary further list of endemic life, animals found there and only there, and in the 1980s the endemic birds were barely known. There were dozens of them, perhaps seventy to one hundred species, which had evolved through the island's long, human-free isolation. Some had already gone, like the elephant bird – the biggest bird that ever lived – and many of the others were on the brink. I photocopied and transcribed more journals and books and wrote up what others had seen and what I could deduce from the labels on the dead birds in Tring and other museums: news of a kind trapped one hundred years before and thousands of miles away. Everything was dead at the point it entered my book, all the individual birds and all but one or two of the people who had seen them; almost everything was *backstory*. There was very little contemporary information from Madagascar and I never went.

In 1986 the book appeared: *The Endemic Birds of Madagascar*. My life shifted and when, decades later, I went to the island for the first time I'd forgotten almost everything that was in it, until I started seeing some of the birds I had written about and they all assembled themselves before me and then a whole island got up from its long sleep and walked about.

I went to Madagascar with Claire and our friend Callan because he had heard a strange bird-call at night in a forest. All three of us were keen to find out what it was. Claire and Callan are scientific ornithologists and field naturalists. Callan is also a wildlife tour leader. They were eager to solve a mystery: was the call the previously undescribed sound of a little known bird? I knew less but hoped for more: could it be an unknown bird making the unknown call? Might there be a new species to be found?

I wasn't much help. At three in the morning in the rainforest on the Masoala Peninsula, everything was in the dark and beyond me. I stumbled under the scattered starlight and moon shadow. 'Recording!' Callan said, to keep me quiet, from the head of our little Indian-file of three. He aimed his microphone at the dark and switched the recorder on.

I was a stranger to the jungle, and my skills as expedition sound-recordist were failing: for the last few months I had been either deaf in my right ear, my hearing near flannelled to nought, or intermittently tortured in the same place by high-energy tinnitus, cruelly appropriate in the forest, like cicadas operating a circular woodsaw within millimetres of my seat of being. A while back, I had managed to hear our target bird, but I hadn't been able to place the call or record it usefully. Callan had sensibly commandeered the kit.

Behind us, out of sight, the black noise of the night sea broke on a sandy shore, while ahead of us, tall trees rose up blacker than anything and thick with themselves, closing down the path to a dark zero, as if it were no more than a hole in the ground.

There are, though, different ways of being absent. Dark needn't mean empty. A black screen is not the same as a blank screen. The night forest might be either or neither or both. A bird in it – the bird we wanted out – might be known but not there, or there but unknown or, more simply, there in the dark but quiet or sleeping (as I wished to be) like a log among the silent commotion of the trees.

*

On two previous visits to these trees in the north-east of Madagascar, under a full moon on two December nights, Callan had heard and then seen a bird he didn't recognise. Initially he wasn't even sure that the sound he had heard was a bird. A smothered forest stirs in its night-sleep, quilted by the dark, and it occasionally breathes out its dreams in different voices. The quiet *k-tick* Callan heard might have been a frog. But a bird flew towards him and the call came again and accelerated into a song, a liquid-metal trill. He knew then that it was a species of nightjar.

Nightjars are among the least knowable of birds. Configured to hide abroad in both light and dark, they live in the night and look like it, while during the day they sleep and then look like that too – like sleep embodied. Separating them from either night or day requires some juju to answer their voodoo. If you stumble on a nightjar roosting on the ground or guarding its eggs, or young in its near-subliminal nest-scrape, it will flick hurriedly away in a flying gangle, hating the tombstone glare of the day, often slumping back to the ground as if injured, so as to distract you from its home patch. If you manage to peel one free from its camouflage, and spot it without flushing it from its roosting branch or nest scar, it will remain fixed and immobile, stopping time – absorbing it, it seems – through the ancient clothes it appears to be wearing and the great black globes of its dozing eyes. Jinxed and stunned, you pick yourself up and move creakily away, a deadbeat lost in nightjar time.

They are better caught at night, and traditionally with a handkerchief. Many nightjars have white patches on their tail and wings and these spots, the simplest marks on the birds, function as badges for their needs. Males display by showing their whites like so many moons thrown into the night sky. Because of this I saw my first ever nightjar on a summer heath at Thursley Common in Surrey, when I was nine and my father took me to a nightjar evening run by the local bird club. Apart from him and me, most of the participants wore headscarves, with our elderly leader in a deerstalker and tweeds. We arrived a little late and hurried to catch up with the group as it shuffled to a halt. The common had begun to purr with nightjars. One churring male had set another off and the whole heath reverberated. Our leader stepped ahead and took a folded handkerchief from his breast pocket. He flapped it open and waved it at the night, like a surrendering soldier, its small white flag brushing his head. Suddenly, a nightjar was there, loudly snapping its wings above its body with a feathered whiplash – a male drawn out of the night by the white of the handkerchief – and flashing his rival white tail- and wing-spots just above the old man.

Aside from that one-time Morris dance, nothing that I know of nightjars is obvious. All of the world's 135 or so species are tailored somewhere between a fast owl and a stuttering moth – long sharp wings, wide-mouthed faces, blackcurrant eyes – and they are silently assembled, as are many night things, from old fabrics woven and rewoven, stitched and unstitched, worn, patched and appliquéd. If an owl is a worked bag of leafy air, a nightjar is a dusty carpet that has absorbed into its pattern the tread of that which has been trodden down into it, until it cannot be said what is dirt and what is design.

Around the world, wherever there are nights warm enough to eat outdoors, there are nightjars that look like this. Each species runs rigs, genius variations, out of these concealing dark materials, browns and greys and blacks, taking on the tone of the particular night sky they fly beneath and the grain of the earth where they rest and breed. All of them spring alive from catalogues of camouflage like flying carpets or moving earth, each is its own chthonic dhurrie.

And from these strange elsewhere places, underlands as deep-cut as between those consonants, come the nightjars' calls and songs. They are by-catch trawled from beyond the sound horizon, earth-purrs and night-slips, netted to the surface and nearly brought into the sonic range of the world's ear, but felt as often in the chest as in the head: calls like chips off scattered stars, shared with cicadas and crickets and frogs, and songs like pre-industrial pumps that sound the same as the ooze they pump, motors blurred with the motored, draining uphill old oils made of still older wood and soil, leaf-mould opening a throat.

And because they are the night's things and we are not, nightjar kinesis appears as weird as their crypsis and as thrown as their voices. They move too fast or not at all. By day they play the part of a log. In those hours, they are friends with the forest floor or a rock's flat face. They sit tight, so still that lichen might advance over them. Once, in Crimea, I watched a mosquito land on the wet eyeball of a grounded nightjar and it didn't blink. By night they work as a headache. Moths are what they mostly want to catch and eat, and the birds fly after them, their wing-scythes and hawking jinks filleting the dark sky. Under unclouded moonlight these actions might be computable. But crossing a torch-beam or a headlight directed upward into an overcast sky, the same flights get chopped into a bat-flit origami – the sky is being shredded as well as folded, the birds appearing all too quickly on their own stage and exiting at the same moment, each its own chronic hurry.

Such words as these I sent after the facts, trying to chase them down, in an absent-minded fugue, and so I missed the path in Masoala and bumped into Claire and Callan. Behind these professionals, at the end of the line, I had fashioned a creature bogged with my own guff. It was just one way of putting it. Claire and Callan would have others. But, for now, we'll do something altogether different – press a button and start a machine. Stop recording the song of the earth and begin playing back. Sound out a bird with its own tune. Call it in. It's juju time.

Two years before, Callan had recorded what he heard and replayed it. The unknown bird leapt at the sound of itself and flew to an overhead tree but moved off before Callan could focus his camera. He saw it for only a few seconds in the dark. He could tell it was a nightjar, but which nightjar?

Madagascar has two known species – the Madagascar and the collared. Both are endemic and found nowhere else.

Both are known to be at home in Masoala, and both have white in their plumage. Like many around the world, the Madagascar incarnation obeys the family template of mixed browns with white flashes on its wings and tail. Its song also falls within nightjar-speak, and is described in one field guide as being like a ping-pong ball bouncing on a table and in another like a marble bouncing on a stone floor. I'd heard some before we got to Masoala and transcribed it in my notebook as a creaky handle on a labouring old pump.

The collared nightjar is stranger than the Madagascar and not so sporting. Its name comes from a distinctive buffish necklace that, combined with a rufous scarf around the nape, gives the bird a more hunched and bigger-headed appearance than many other nightjars. It is a member of a less textbook family too, only recently untangled from other relatives, and the male has white in its tail but not in its wings. Unusually for a nightjar, it prefers to nest off the ground in the soft root mass of bird's-nest ferns that grow as epiphytes on forest trees. Not so widely distributed on the island as the Madagascar, the collared is generally less familiar. Its voice has never been described. One book notes this absence as an 'abiding mystery'.

Callan had heard something new in Masoala and now Claire and I did too. A full moon had risen and made its way through the top branches of the dark wood. After a blast of amplified playback, a digital surge meeting insect hiss and the odd clock tock of the recorded bird, something real called back and the bird was there, all at once, flying towards the moon and up into a tree, landing on a bare branch. Callan fired his camera.

★

Evolution isn't over, although most of us carry on as if it has finished – as if its discovery, the auditing of its accounts, locked it down forever. Knowledge of our own inevitable extinction and of our stupefaction when contemplating deep time skews our thinking. Species are coming into existence as much as they ever were, but unscientific people like me, preoccupied with our *Weltschmerz*, think of species as mostly disappearing, and mostly by our hand. When Claire and I were courting – I waving handkerchiefs, she weighing her options – Claire said to me that the one thing she knew to be true about the world was that natural selection operates on every living thing and that it is happening still. I knew then that I wanted to marry her, even if I didn't quite get her point. With my white flag I was waving *hello* and *I surrender* at the same time.

A month in Madagascar might wake you up in a similar way. Our little and late understanding of what makes a species has been played out and projected across Madagascar's extraordinary wealth of endemic animals: a fauna that includes most of the world's chameleons, the entire clan-membership of bird families like vangas, couas, mesites, and ground-rollers, and all the mammals that are tenrecs and lemurs.

The reason why nowhere else has such things has been debated as long as Madagascar has been known to the rest of the world. The roc, the giant bird that carried off Sinbad in *One Thousand and One Nights*, may well have been hatched from early accounts of elephant birds on the island. Marco Polo mentioned these stories. (He is also said to have given Madagascar its name, though he never saw it, having misheard or misunderstood the word Mogadishu.)

These days some scientists believe it is possible that all the lemurs of Madagascar might constitute a clade, sharing a single ancestor – a pregnant female – who was unwittingly translocated and arrived, as if shipwrecked, on a raft of vegetation washed hundreds of miles across the sea from the African continent. In 2013, there were thought to be ninety-eight species of lemur on the island, whereas in 2003 just forty were reported. Time hurries or dawdles on Madagascar like anywhere else: no new lemurs appear to have come into life in that decade, but during those years their bodies – like the gulls of the northern hemisphere – were turned about by our science: picked up, looked at, redescribed, put down.

Imagine if somewhere east of the Masoala Peninsula, further out into the Indian Ocean, on the islands along the Mascarene Ridge, a dodo was found quietly getting on with life, like an unsurrendered Japanese soldier walking out of the forests of Burma. What would that do to our minds? How might we see the dodo then? Like Elvis? As a Lazarus taxon?

Absence tells differently from presence. All the descriptions we have of dodos, all the words spoken and written by those who saw them, are now hundreds of years old. The bird stiffens with age and dies in our heads with the language. All extinctions and losses are like this. They die twice over. Why do my great grandparents occur to me in black and white? Because they survive purely through the monochrome technology of their time. Why do dinosaurs only have scientific names? Because there were no Maori to call them *moa*, no Dutch sailors to call them *dodo*.

What happened, Borges asked in his parable 'The Witness', when the last person died who had once seen Christ alive? As the lost disappear from the earth and from our minds, dying twice, so all discoveries live two times over as they come into known existence through language. It was like this in the middle of the night in Masoala when we three spoke of the bird that we had just heard and seen, as no one ever before us would have or could have, for no one who had previously looked or listened had been able to match the sound to its source.

Species must be brought to life with a name. Dr Frankenstein doesn't christen his creature, and as a telling consequence his creation is often misremembered as having his name. Mary Shelley wasn't a taxonomist; neither perhaps did she believe that it was Adam's proper task to name the animals in Eden. Yet without those words, we're lost and the living thing invisible.

On Madagascar new species are being discovered in two ways. There are animals being found that have never been seen before by naturalists. They are called new because they were previously unknown to science. These seem the truest or purest additions to the list of life: good news for the world. But then there are new species born – like the gulls of the north – from lumping and splitting, thanks largely to laboratory work. The leap in the Madagascar lemur tally comes down to this. Though the truth is being equally served by both sets of findings, it is hard for the second not to seem secondary, to feel more like a revision or an update.

Fencing a species, calling it down, knowing where it might stop and another begin, finding a partner to walk the plank with it into the ark – none of this can be undertaken without a meeting of the objective and the subjective (the described and the describer, facts and thoughts), together with a fixing of time (the concatenation of years frozen in a snapshot).

In life, all species are in motion and on their way elsewhere. Nothing sits still on the forest floor forever. A new named species is made out of fieldnotes (an attempt to describe the way things *are*) and also the invention of some sort of holding measure that will do for now (an understanding of how things *seem*). It is not easy. Blood must almost always be shed. A new species is still, most commonly, defined by a dead body.

Any new species may well have been there as long as any other. But named and raised into our knowledge, the new – however old it is in world time – will now add its voice to the chorus of the known and bay with it at the dark. Madagascar has announced this more loudly than many places and, if your hearing is acute, you might begin to catch its call – the mixed sounds of the night and of the known. After two weeks travelling from semi-desert in the south-west to north-eastern rainforest, Claire said to me that although she had known about adaptive radiation since she studied at university, before we got to Madagascar she had never felt it operating in front of her.

On our travels we saw vangas across the island that were taking the role of shrikes in their absence, and others from the same family that have become the equivalent of flycatchers, or nuthatches, or wood-hoopoes, or babblers.

On one daytime walk in the forest at Masoala, a mixed bird party passed through the canopy above us and was lit up with several species of vanga (blue, Chabert's, rufous, red-tailed and Bernier's) busy being themselves as well as ghosting the unrelated birds they'd taken after. Once the party had moved on and the forest fell quiet, its beautifully disturbed air stilled, we sat under the trees. Claire told me about phylogenetics and the branching of lines while I filled her in on the Crummles in *Nicholas Nickleby*, the family of theatrical entertainers who take on all the parts and do all the voices, reaching into a pantomime basket filled with every bill and beak, cloak and cape, gown and mask.

Shifting those bodies, raising them into life in the light, bringing them and their facts into the wider world, was our goal. So, in the night forest at Masoala, under the nightjar tree, as soon as he had taken the photographs, though the bird was still there above us to be looked at, Callan scanned the screen of his camera to see what it showed – to check what he'd caught, to ask if it was enough.

*

The nightjar we watched on the branch had come to the replayed call of the first bird Callan had heard and then seen. What we could see of this new bird on the camera screen clinched it. We had discovered the call of the collared nightjar – nothing more but nothing less. We hadn't found a new bird, but we had untangled the sound of a hidden singer from the night-noise and had pinned it onto a mass of feathers.

I'd come to Madagascar to try to colour in the faded bodies I knew from my book-making days at Tring. To see the birds in the forests was to bring to life what I had only known as dead. But I'd also hoped to find something new, to fetch something out of the dark. Although I was less scientifically serious about this than Claire or Callan, I seem to have minded more when we didn't. They were excited with the new data. The call and the song of the collared nightjar are now known. But I surprised myself by how disappointed I felt. I hadn't thought I was a collector. I've known for years that no true claim can be staked. The world is not ours to get or to spend. But to be among the first eyes and ears to have knowingly seen and heard a bird that no one had seen and heard before – wouldn't that be something? Wouldn't that stitch you to another's heart forever? To be a finder would be as a close to being a maker as is possible. And if you love the world why wouldn't you want more of it?

Once identified, the bird's song sounded sad. I infected it as it came to my damaged ears. We had collared the nightjar, but doing so opened up to me the countless years when the birds had called and sung and only their own kind had known what was being said. Of course, the human villagers of Masoala, walking the woods at the end of their day, may well have heard and recognised the nightjar for what it is and was. And, of course, the bird cared neither way. But hearing the call and being among the very few people who then knew what there was to be heard, and who had thereby put it into conscious thought, made the sound sad. Loneliness came off it.

★

In the last minutes of our last night at Masoala we had our best view of our bird. It gets light at about four-thirty in the morning there in December. We'd been out on the trail all night. Already the day was opening high and fast when suddenly right next to the cabin where Callan was staying, on a bare branch towards the top of a spreading tree, came first one and then another collared nightjar, calling agitatedly in response to our canned sounds. We didn't need torches to light the branch – daylight was doing that. The birds seemed exposed now, tricked by our version of their music and caught out in the coming light.

A pale blue was pressing through the grey overhead, and from another tree a bird called, a bulbul, the first of the morning, and everything of the dark, the night and the nightbirds, seemed at once old and uncoloured, finished and no longer wanted. It was all put away in a trice and the dawn won as easily as that. And, as if they knew it, the nightjars got up and flew fast from the branch, hugging the dark comfort of leaves, and then twisted through a small shaded opening and were gone.

The Family Line

Back in Britain, I laboured for a time under a personal *dustology*. A quarrying of some of my own internal geology was called for. I had grown a kidney stone until it announced itself like a child that would not be born. I was first infected, then winded by its cannonball lurk. It was declared too big to pass and had to be temporarily sidelined with a stent. Awaiting granulation by laser, I lived for weeks beneath the stone's moony shadow, peeing bloody and earthy vintages with wincing trepidation, walking with difficulty, and, latterly, stooping in pain as I stood. On good days, I felt like the Little Prince leaning nervously over his own planet; on bad ones, like Sisyphus wedded to his rock. The medical talk I longed for was of a divorce: the smashing of the stone to something gritty that might exit my body as gravel or sand or – better still – dust. Finally, after weeks of horizontal waiting, I met Mr Timoney, an expert surgeon. On a brief trip inside me, he aimed his laser and was able to both split-up and collect. There followed twenty-four hours of razorblade pain as the last of the silt inched from me, before my life was wonderfully restored and my story was no longer solely my stone's.

As I left the hospital after my last stay, I looked through its skylights up at a blue day. Confined to my room for months, my horizons had shrunk and I had sorely missed the sky. For ages I had hardly seen a gull, but there were some up there – lesser black-backed gulls planing overhead. The first summer the new hospital was built, some of these birds had attempted to breed on its flat roofs. Hawk-shaped kites and a shouting scarecrow hadn't worked. Worse still, the gulls discovered the loose stones laid on the roof to help rainwater drainage. They were drawn to them and, quickly, fifty of the roof panels were shattered when the birds dropped the pebbles onto the glass. The hospital employed a pest-control company who destroyed the gulls' nests, and flew Harris's hawks over the site to scare off any lingerers. The company declared victory over what they called the 'anti-social' birds but, three years later, as I walked, stone-free, under the atrium ceiling, the gulls overhead jinked and jumped as they flew above several new shattered glass panels.

*

As I recuperated, my walks to the seafront at Minehead, after hours spent with my parents in their house, became more common. My father, having nursed my mother, got ill himself and had become increasingly static. I drove from Bristol to change light bulbs, cut the grass, jettison mouldy gooseberries from the fridge, and organise trips to the local dump, a strictly no food waste place with metal skips for sorted recyclables. In the house, stuff had built up that needed clearing. But getting rid of your life is hard. My mother, a would-be life-laundress, offered me my pick of the paintings on their walls and then found reasons to deny all of my choices. My father sanctioned my clearing various bin bags of plastic tubs and boxes, only for me to see his bent frame falling slowly into one of the skips as he sought to retrieve a favoured yoghurt pot which he had emphatically not authorised for extinction.

Dad's spine had jackknifed and his heart was weakening. Mum was still lame. Baby, their ginger and white cat, got ill as well and had to be shaved. Dad took to carrying the cat around by wedging it between his thighs and stomach. He walked from the front room to the kitchen looking like a veteran miner, stooping from his shift, blinking into the light, and cradling a sack of golden nuggets, the last haul of his life. I topped up the seed and fat balls on the bird table and asked about gulls in the garden. There are herring gulls working the town from overhead all day long and one – both my parents insisted it was just one – came down most mornings onto their pocket-sized lawn and browsed freely. Baby wasn't eating well and recently Gilbert the gull had been enjoying a lot of old cat food. A stale doughnut, put out just that morning, was probably snaffled by him as well. He waited on the roof above their bedroom and they often heard his claws on the tiles and the gutter.

We tussled over what I might take on my next dump trip. Dad gave up some newspapers he had vetted, but others that hadn't been fully read still hummed for him with latent energy and had, therefore, to be kept. There was a skull of yellowing paper that could have been a wasps' nest. I put my fingers through it and fished out a booklet on how to pass the eleven-plus. I'm fifty-seven and my children are in their twenties, my sister and her boys are not far behind us. The exam disappeared from most of England in the 1970s. The booklet remains on my dad's shelves.

Even as I wonder what world is to be won by this, I know that in my own bat cave in Bristol I am the author of similar madnesses, comparable hoardings for unlikely tomorrows. There are shelves but there are a lot of books beyond shelving, there are vinyl records and record players, stuffed birds and old bird reports, there are old radio tapes and old radios, there are towers of CDs, there are works on foolscap, roneo and video, there is a mothy rug, a box of feathers, some baskets of stones, another of blown ostrich eggs, there are mantelpieces decked with whalebones and bird skulls, there is half a table of fulgurites, gnarled glass fingers of lightning-struck Saharan sand, there are rings from dead birds. And, until it died itself, I had a frozen-up freezer; which melted onto my floor and out slid a grey mullet hooked from the muddy tide of the Wash, along with some defrosting peas, a barn owl's wing sliced from a roadkill, an ice-cream pot filled with ancient mulligatawny, and the disassembled paw of an aardvark found on a dirt road in the Northern Cape.

Wasps make magnificent homes for themselves but every winter all of the builders die. Many of us know this and its wider implications, but most of us also live as if this weren't true.

At my parents' house, I opened a cupboard in a lean-to that had once been a car porch. It was full of water-stained stationery, memo pads, notebooks and some rusted paper clips. None had been used and all were now useless. Dad shouted 'No!' and did his best to hurry towards me. Not even the yoghurt-pot rescue had exercised him as much. I stopped my bagging. As he walked away, I could hear effort in his chest, a wet wind of muddied air, like the sea's breath at a shore or, as Robin Robertson says in his poem 'A Seagull Murmur', 'the mewling sound of a leaking heart.'

There was wine at lunch and cheesecake. The heating was working overtime. Dad remembered his National Service duties in Plymouth Sound, where he fired twelve-pounder guns at a radio-controlled target boat called the Queen Gull. I'd heard of the gun batteries before but never of their marks. After coffee, both parents dozed in their armchairs. I stood up to get some air and walked to the front. It was a blowy day. The first sand martins of the spring had been seen in half-a-dozen places across southern England, pioneers of the incoming season. But though today's wind had fetched out of the south-west, winter was in it still. My eyes swam with tears. It was one of those days when the Atlantic bullies the Bristol Channel, requiring it to awake its faith, up its game, and audition for *sea*. Though the washed out wet-waste of half of England and Wales kept the water brown, like potter's slip, there were deep troughs and white-crowned waves and everything *streamed*.

Herring and lesser black-backed gulls spun out of the cloud base and came down like squalls, bright lit by spears of sunlight between broad shawls of grey rain. The Minehead Chamber of Commerce had posted a sign with a picture of a herring gull outside the mini-golf café saying Please Don't Feed Me. I crouched beneath it to watch the birds. The tide was out, the beach wide. In the first water, three sickly herring gulls were washing themselves. Two of them had one leg that hung uselessly. All their steps were hops. The third had wing feathers that had been half eaten away. Those that remained were stained – it looked like it had fallen into something corrosive. It was drinking heavily.

Further out, at the limit of my binoculars' power, many more gulls splashed and washed in the spreading tongue of Minehead drain water where it debouched into the brown fringe of the sea. There was quite a crowd: five hundred herring gulls, one hundred common gulls and a dozen each of lesser and great black-backed gulls. I thought of the holy dirty water of the Ganges – its lower reaches declared a 'living entity' in 2017 and, shortly after, reported 'dead' – and of the outlets, just up the estuary, of the nuclear power plants, Hinkley A, B, and one day C.

One bird among the many detained me. The stormy evening was floodlighting everything where the sea began, and one gull shone out still more, even though it was a third of a mile away. A ghost gull – the colour of dirty ice or wood ashes. It was smaller than the herrings and more delicately built with longer wings. Its body was mucky but overall it appeared much brighter than any other gull. It was like an ice-light or snow-lantern on the shore. It primaries were the whitest part of it. No moon or spoons or mirrors – just white. It looked like it was thawing. I knew it at once as an Iceland gull, only my third ever and the first I had found alone. A full adult would have been whiter still. This was an immature, but I was too far away to properly age it by calculating the full extent of youthful dirt that it wore. I loved it immediately nonetheless, and I knew, as I watched its northern light dimming through the Somerset dusk, that it would end this book.

Bibliography

All titles published in London unless stated

Allen, David Elliston, *Books and Naturalists*, Collins, 2010

Allen, David Elliston, *The Naturalist in Britain*, Penguin, Harmondsworth, 1978

Alter, Robert, *The Five Books of Moses*, W. W. Norton, 2004

Armitage, Simon, and Dee, Tim, (eds.), *The Poetry of Birds*, Penguin, 2009

Armstrong, Edward A., *The Folklore of Birds*, Houghton Mifflin, Boston, 1959

Armstrong, Edward A., *Shakespeare's Imagination*, Lindsay Drummond, 1946

Bach, Richard, *Jonathan Livingston Seagull*, Harper Element, 2003, first published 1972

Balmer, D. E., Gillings, S., Caffrey, B. J., Swann, R. L., Downie, I.S. and Fuller, R. J., *Bird Atlas 2007–11*, BTO Books, Thetford, 2013

Beckett, Samuel, *The Complete Dramatic Works*, Faber, 1986

Berger, Alan, *Drosscape*, Princeton Architectural Press, New York, 2006

The Bible, *Authorized King James Version*, OUP, Oxford, 1997

Bircham, Peter, *A History of Ornithology*, Collins, 2007

Birkhead, Tim, Jo Wimpenny and Montgomerie, Bob, *Ten Thousand Birds – Ornithology since Darwin*, Princeton University Press, Princeton and Oxford, 2014

Blake, William, *Complete Writings*, edited by Geoffrey Keynes, Oxford University Press, 1969

Brown, Andy, and Grice, Phil, *Birds in England*, T. & A. D. Poyser, 2005

Borges, Jorge Louis, *The Book of Imaginary Beings*, translated by Norman Thomas di Giovanni, Penguin, Harmondsworth, 1987

Borges, Jorge Louis, *Collected Fictions*, translated by Andrew Hurley, Penguin, 1998

Borges, Jorge Louis, *Labyrinths*, edited by Donald A. Yates and James E, Irby, Penguin, Harmondsworth, 1985

Boyd, A. W., *A Country Parish*, Collins, 1951

Browne, Thomas, *Notes and Letters on the Natural History of Norfolk*, Jarrold & Sons, Norwich, n.d.

Browne, Thomas, *Religio Medici and Hydriotaphia, or Urne-Buriall*, edited by Stephen Greenblatt and Ramie Targoff, New York Review Books, New York, 2012, first published 1643 and 1658

Burrows, Tim, *The Only Grave is Essex*, Guardian, 25 October 2016

Campbell, Bruce, and Lack, Elizabeth, *A Dictionary of Birds*, T. & A. D. Poyser, Calton, 1985

Campkin, Ben, and Cox, Rosie, eds., *Dirt*, I. B. Taurus, 2012

Celan, Paul, *Threadsuns*, translated by Pierre Joris, Green Integer, Copenhagen and Los Angeles, 2005

Chekhov, Anton, *The Seagull*, translated by Tom Stoppard, Faber, 1997

Cioran, E. M., *A Short History of Decay*, translated by Richard Howard, Penguin, 2010, first published 1949

Clare, John, *John Clare's Birds*, edited by Eric Robinson and Richard Fitter, Oxford University Press, Oxford, 1982

Clare, John, *The Oxford Authors: John Clare*, edited by Eric Robinson and David Powell, Oxford, 1984

Cleere, Nigel, *Nightjars of the World*, Wild Guides, Old Basing, 2010

Cocker, Mark, *Birds Britannica*, Chatto & Windus, 2005

Cocker, Mark, *Birds and People*, Jonathan Cape, 2013

Cocker, Mark, *Richard Meinertzhagen*, Secker & Warburg, 1989

Coetzee, J. M., *Age of Iron*, Penguin, 2000

Coleridge, S. T., *Selected Poems*, edited by Richard Holmes, HarperCollins, 1996

Collar, N. J. and Stuart, S. N., *Threatened Birds of Africa and Related Islands, The ICBP/IUCN Red Data Book, Part 1*, International Council for Bird Preservation, Cambridge, 1985

Collett, Anthony, *The Changing Face of England*, Nisbet, 1926

Collett, Anthony, *The Heart of a Bird*, Nisbet, 1927

Collinson, J. Martin, *CSI: Birding – DNA-based identification of birds*, British Birds, 110, 2017, 8-26

Collinson, J. Martin, Parkin, David T., Knox, Alan G., Sangster, George, and Svensson, Lars, *Species boundaries in the Herring and Lesser Black-backed Gull complex*, British Birds 101, 2008, 340-363

Coward, T. A., *The Birds of the British Isles*, Warne, 1950

Cox, Rosie, George, Rose, Horne, R. H., Nagle, Robin, Pisani, Elizabeth, Ralph, Brian, and Smith, Virginia, *Dirt*, Profile, 2011

Cramp, Stanley, chief editor, *Handbook of the Birds of Europe, the Middle East and North Africa, The Birds of the Western Palearctic, volume III, waders to gulls*, Oxford University Press, Oxford, 1983

Cullen, Esther, *Adaptations in the Kittiwake to Cliff-Nesting*, Ibis, 99, 1957, 275-302

Dee, T. J., *The Endemic Birds of Madagascar*, International Council for Bird Preservation, Cambridge, 1986

Dekkers, Midas, *The Way of All Flesh*, translated by Sherry Marx-Macdonald, Harvill, 2000

Deleuze, Gilles, and Guattari, Félix, *A Thousand Plateaus*, translated by Brian Massumi, University of Minnesota Press, Minneapolis,1987

Dennis, Mike, *Yellow-legged Herring Gulls in Essex*, British Birds, 85, 1992, 246

Dennis, M. K., *Yellow-legged Gulls along the River Thames in Essex*, British Birds, 88, 1995, 8-14

Desilvey, Caitlin, *Curated Decay*, University of Minnesota Press, Minneapolis, 2017

Dickens, Charles, *Bleak House*, edited by Nicola Bradbury, Penguin, 2003, first published 1853

Dickens, Charles, *Nicholas Nickleby*, edited by Mark Ford, Penguin, 1999, first published 1838-1839

Dickens, Charles, *Our Mutual Friend*, edited by Adrian Poole, Penguin, 1997, first published 1865

Douglas, Mary, *Purity and Danger,* Routledge & Kegan Paul, 1966, revised 2002

Eliot, George, *Middlemarch*, ed. by Rosemary Ashton, Penguin, 2003

Eliot, T. S., *The Poems*, Volume I, edited by Christopher Ricks and Jim McCue, Faber and Faber, 2015

Empson, William, *The Complete Poems,* Allen Lane, 2000

Ferns, P. N., and Ross-Smith, V. H., *Function of Coloured Bill Tips, Stripes, and Spots in Breeding Gulls*, Marine Ornithology, 2009, 37, 85-92

Fisher, James, *The Shell Bird Book*, Ebury Press, 1966

Fisher, James, and Lockley, R. M., *Seabirds*, Collins, 1954

Fitzgerald, F. Scott, *The Great Gatsby*, Penguin 2000, first published 1925

Foucault, Michel, *The Order of Things*, Routledge, 2002, first published 1966

Garfield, Brian, *The Meinertzhagen Mystery: The Life and Legend of a Colossal Fraud*, Potomac, 2007

Garner, Martin, and Quinn, David, *Identification of Yellow-legged Gulls in Britain*, British Birds 90, 1997, 25-62

Garner, Martin, Quinn, David, and Glover, Brian, *Identification of Yellow-legged Gulls in Britain. Part 2*, British Birds 90, 1997, 369-383

Gibbins, Chris, Neubauer, Grzegorz, and Small, Brian J., *Identification of Caspian Gull Part 2: phenotypic variability and the field characteristics of hybrids*, British Birds 104, 2011, 702-742

Gibbins, Chris, Small, Brian J., and Sweeney, John, *Identification of Caspian Gull Part 1: typical birds*, British Birds 103, 2010, 142–183

Gissen, David, *Subnature*, Princeton Architectural Press, New York, 2009

Goodman, Steven M., and Benstead, Jonathan P., eds., *The Natural History of Madagascar*, University of Chicago Press, Chicago, 2003

Grant, P. J., *Gulls: A Guide to Identification*, T. & A. D. Poyser, Calton, 1982

Greenlaw, Lavinia, *Minsk*, Faber and Faber, 2003

Greenoak, Francesca, *All the Birds of the Air*, Andre Deutsch, 1979

Greenwood, Jeremy, *Getting our Lists in Order*, British Birds, 110, 2017, 250-251

Gurney, J. H., *Early Annals of Ornithology*, H. F. & G. Witherby, 1921

Hanlon, James, *UK500: Birding in the Fast Lane*, Brambleby, Harpenden, 2006

Harrison, Stephen, Pile, Steve, and Thrift, Nigel, eds., *Patterned Ground*, Reaktion, 2004

Harting, James Edmund, *The Ornithology of Shakespeare*, Gresham Books, Old Woking, 1978, first published 1864

Hawkins, A. F. A. and Morris, Pete, *Birds of Madagascar*, Pica, 1998

Hawkins, Frank, Safford, Roger, and Skerritt, Adrian, *Birds of Madagascar and the Indian Ocean Islands*, Helm, 2015

Hecht, Anthony, *The Transparent Man*, Oxford University Press, Oxford, 1991

Holloway, Simon, *The Historical Atlas of Breeding Birds in Britain and Ireland 1875-1900*, T. & A. D. Poyser, 1996

Hoyo, J. del, Elliott, A., and Sargatal, J., eds., *Handbook of the Birds of the World, volume 3, Hoatzin to Auks*, Lynx Edicions, Barcelona, 1996

Hoyo, Josep del, Collar, Nigel J., Christie, David A., Elliott, Andrew, Fishpool, Lincoln D. C., *HBW and BirdLife International Illustrated Checklist of the Birds of the World. Volume 1: Non-passerines*, Lynx Edicions, Barcelona, 2014

Hoyo, Josep del, Collar, Nigel J., Christie, David A., Elliott, Andrew, Fishpool, Lincoln D. C., Boesman, Peter, and Kirwin, Guy M., *HBW and BirdLife International Illustrated Checklist of the Birds of the World. Volume 2: Passerines*, Lynx Edicions, Barcelona, 2016

Hudson, W. H., *Birds in London,* Longmans, Green, and Co., 1898

Hume, R. A., *Variations in Herring Gulls at a Midland roost*, British Birds, 71, 1978, 338-343

Keats, John, *John Keats – The Oxford Authors*, edited by Elizabeth Cook, Oxford University Press, Oxford 1990

King, Clive, *Stig of the Dump*, Penguin, Harmondsworth, 1963

Kruuk, Hans, *Niko's Nature*, Oxford University Press, Oxford, 2003

Larkin, Philip, *Collected Poems*, Marvell Press with Faber and Faber, 1990

Logan, William Bryant, *Dirt*, W. W. Norton, New York, 2007

Marx, Karl, *Capital*, Volume. 1, translated by Ben Fowkes,
    Penguin, 2004

Maurier, Daphne du, *The Apple Tree*, Victor Gollancz, 1952

Mayhew, Henry, *London Labour and the London Poor, four volumes*,
    Dover, New York, 1968, first published 1861-1862

Mayhew, Henry, *The Unknown Mayhew*, edited by E. P. Thompson
    and Yeo, Eileen, Penguin, Harmondsworth, 1973

Melville, David S., *Yellow-legged Herring Gulls in Essex*, 1973-74,
    British Birds, 84, 1991, 342-343

Mitchell, Dominic, *Slaty-backed Gull in London and Essex: New to
    Britain*, British Birds, 110, 2017, 405-413

Munsterberg, Peggy, ed., *The Penguin Book of Bird Poetry*, Penguin,
    Harmondsworth, 1984

Nicolson, E. M., *Bird-watching in London*, London Natural
    History Society, 1995

Olsen, Klaus Malling, and Larsson, Hans, *Gulls of Europe, Asia and
    North America*, Helm, 2003

Parslow, John, *Breeding Birds of Britain and Ireland – a historical
    survey*, T. & A. D. Poyser, Berkhamsted, 1973

Potter Stephen, and Sargent, Laurens, *Pedigree: the Origins of Words
    from Nature*, Taplinger, New York, 1974

Real, E., Oro, D., Martínez-Abraín, A., Igual, J. M., Bertolero, A.,
    Bosch, M., and Tavecchia, G., *Predictable anthropogenic food
    subsidies, density-dependence and socio-economic factors influence
    breeding investment in a generalist seabird*, J Avian Biol. 2017, 48,
    1462-1470

Robertson, Robin, *Swithering*, Picador, 2006

Rock, Peter, *Urban Gulls: Problems and Solutions*, British Birds, 98,
    2005, 338-355

Rock, Peter, *Urban Gulls. Why Current Control Methods Always Fail*,
    Riv. ital. Orn., Milano, 81, 2013, 58-65

Rock, Peter, and Vaughan, Ian P., *Long-term estimates of adult
    survival rates of urban Herring Gulls Larus argentatus and
    Lesser Black-backed Gulls Larus fuscus*, Ringing & Migration
    2013, 28:1, 21-29

Rock, Peter, Camphuysen, C. J., Shamoun-Baranes, Judy, Ross-Smith, Viola H., and Vaughan, Ian P., *Results from the first GPS tracking of roof-nesting Herring Gulls Larus argentatus in the UK*, Ringing & Migration 2016, 31:1, 47-62

Selous, Edmund, *The Bird Watcher in the Shetlands with some Notes on Seals and Digressions*, J. M. Dent, 1905

Selous, Edmund, *Evolution of Habit in Birds*, Constable, 1933

Selous, Edmund, *Thought Transference (or What?) in Birds*, Constable, 1931

Shakespeare, William, *Complete Works*, edited by Jonathan Bate and Eric Rasmussen, Macmillan, Basingstoke, 2007

Shrubb, Michael, *Feasting, Fowling and Feathers*, Poyser, 2013

Sigurdsson, B. D., and Magnusson, B., *Effects of seagulls on ecosystem respiration, soil nitrogen and vegetation cover on a pristine volcanic island, Surtsey, Iceland*, Biogeosciences, 2010, 7, 883–891

Smith, Cecil, *The Birds of Somersetshire*, Van Voorst, 1869

Steedman, Carolyn, *Dust*, Manchester University Press, Manchester, 2001

Stoddart, Andy, and McInerny, Christopher, *The 'Azorean Yellow-legged Gull' in Britain*, British Birds, 110, 2017, 666-674

Sullivan, Robert, *The Meadowlands*, Granta, 2006

Tauler-Ametller, H., Hernández-Matías, A. Pretus, J. LL., and Real, J., *Landfills determine the distribution of an expanding breeding population of the endangered Egyptian Vulture Neophronpercnopterus*, Ibis doi: 10.1111/ibi.12495

Tinbergen, Niko, *Curious Naturalists*, Penguin, 1974

Tinbergen, Niko, *The Herring Gull's World*, Collins, 1953

Tinbergen, Niko, *The Study of Instinct*, Oxford, Oxford, 1952

Tinbergen, N., and Perdeck, A. C., *On the stimulus situation releasing and the begging response in newly hatched Herring Gull chick*, Behaviour 3, 1950, 1-39.

Tristram, H. B., *The Natural History of the Bible: being a review of the physical geography, geology and meteorology of the Holy Land; with a description of every animal and plant mentioned in Holy Scripture*, Society for Promoting Christian Knowledge, 1868

Turner, William, *Turner on Birds*, edited by A. H. Evans, Cambridge University Press, Cambridge, 1903, first published 1544

Vinicombe, K. E., *Ring-billed Gulls in Britain and Ireland*, British Birds, 78, 1985, 327-337

Virgil, *The Georgics*, translated by C. Day Lewis, Jonathan Cape, 1940

Walker, David, *Status of Yellow-legged Gull at Dungeness, Kent*, British Birds, 88, 1995, 5-7

Wentworth Day, James, *Book of Essex*, Egon, Letchworth, 1979

Wernham, Chris, Toms, Mike, Marchant, John, Clark, Jacquie, Siriwardena, Gavin, and Baillie, Stephen, *The Migration Atlas: Movements of the Birds of Britain and Ireland*, T. and A. D. Poyser, 2002

White, Gilbert, *The Journals, three volumes*, edited by Francesca Greenoak, Century, 1986-1989

White, Gilbert, *The Natural History of Selborne*, Little Toller, Dorset, 2014, first published 1778-1779

Whitman, Walt, *Leaves of Grass and Other Writings*, edited by Michael Moon, Norton, New York, 2002

Wordsworth, William, *Selected Poetry*, edited by Nicholas Roe, Penguin, 1992

Yalden, D. W. and Albarella, U., *The History of British Birds*, Oxford University Press, Oxford, 2009

Yarrell, William, *A History of British Birds*, fourth edition in four volumes, Van Voorst, 1871-1875

Acknowledgements

I have watched gulls myself for fifty years or so, but never as keenly as the men who let me watch them watch, and who talked to me as they did. I am very grateful to them. Without their words there would be no book. I spoke to one gull-woman too and to my wife and two other friends on related bird business; I thank them all equally.

These people are effectively my co-authors: their words and wisdom make up much of what is here: Peter Rock, Paul Roper, James Hanlon, Mark Ward, Dominic Mitchell, Chris Gibbins, Viola Ross-Smith, and Nigel Collar. All gave up days of their lives, generously agreeing to talk to me and, when possible, to show me gulls while doing so. Everyone was friendly and forthcoming and no one laughed me off-site even though my questions and my gull-stupidity may well have deserved it. All tolerated me waving a tape machine at them at the same time (two even allowed me to do this more than once when, after a mental blip, I apparently lost – dumped, even – our first recordings).

Equal thanks also go to Greg Poole for both literally drawing this book and for capturing its idea so vividly. Greg has been a friend of mine for longer than almost anyone else mentioned here – it is very happy making for me that we have now worked together.

My thanks also to the North Thames Gull Group and (through them) Veolia UK at Pitsea and to the Natural History Museum at Tring for officially allowing me into their special places.

My wife, Claire Spottiswoode, allowed me into her mind and her country, showing me the gulls of South Africa and drilling me in evolutionary biology and behavioural ecology in Madagascar. She has kept me alive through the last ten years. I love her for it. She is not to blame for any of the scientific idiocy I show here. She also enabled our adventure in Madagascar, where with her (and now my) friend Callan Cohen, who also has my great thanks, I crashed about and got in the way in a forest at night.

Others kindly gave words to this book. Big thanks to Giggsy, Andrew Tongue, Lesley at Temple Meads Station in Bristol, Patrick McGuinness, Russell Munson, Paul Farley (including his permission to publish his gbbg poem), and the daily gull-feed that BirdGuides.com provides. I remember here my late friend Richard Warr and my sick friend Antony Merritt. I am also in great debt to all other living and dead authors whose gull and rubbish and taxonomical words I have worked over and recycled.

My children, Dominic and Lucian; their mother, Stephanie Parker; and my parents, Kate and John Dee – have all helped and indulged me. I've left their company to watch gulls at times and I've dragged them with me at others and have stolen their words and thoughts. None have complained.

My publisher and editor, Adrian Cooper, has been patient, kind and smart. I also thank Gracie Cooper, Graham Shackleton and Jon Woolcott at Little Toller. My agent,

Anna Webber, is very good news; Seren Adams is excellent too. I must also cheer for Mr Koupparis and Mr Timoney and their expert teams at Southmead Hospital (North Bristol NHS) – without them 100 per cent no book.

Almost everyone I spoke to as I wrote had things to say about gulls. This fact lies behind the whole of *Landfill*. My thanks to all and especially to Richard Alwyn, Ken Arnold, Jeff Barrett and *Caught by the River*, Barbara Bender (and the people of Branscombe who talked about their gulls), Tim Birkhead, Julia Blackburn, Tom Bonnett, Rich Bonser (who deserves a chapter here that space prohibited), Hugh Brody, Mark Cocker (who gave me a copy of Malling Olsen and started a thought), Susie Cunningham, Jane Darke, Nick Davies, Andrew Dawes, Paul Dodgson, Sinead English, John Fanshawe, Rose Ferraby, Lavinia Greenlaw (including her permission to have her words beginning everything), Sam Gugliani and *Medicine Unboxed*, Marybeth Hamilton (who put me on to Henry Mayhew), Jeremy Harding, Alexandra Harris, Fraser and Sally Harrison, Philip Hoare, Richard Holmes, Kathleen Jamie, Mark Johnston, Richard Kerridge, Alastair Laurence, Richard Long, Hayden Lorimer and Nancy Wachowich, Richard Mabey and Polly Munro, Andrew MacNeillie and *Archipelago*, Rachel Murray and Nancy Jones (who asked me to speak in Bristol on the subject of *Marine Transgressions*), Adam Nicolson, Chris Parker, Dexter Petley, Christopher Ricks and Judith Aronson, Robin Robertson (including the loan of a line from one of his poems), Fiona Sampson, Jos Smith, Greta Stoddart (who put me onto Funes), Luke Thompson, Keith Vinicombe.

For permissions to quote many thanks to: Curtis Brown Group Ltd for 'Missing Dates' by William Empson and 'The

Birds' by Daphne de Maurier; United Agents for Chekhov's *The Seagull* translated by Tom Stoppard; Faber and Faber for works by Samuel Beckett, T. S. Eliot and Philip Larkin; and Pierre Joris for his translation of Paul Celan.

Some parts of 'Lump' appeared in a different form in my introduction to *Animal, Vegetable, Mineral* (Wellcome Collection, 2016) and some parts of 'Night Watch' appeared in a different form at granta.com. My thanks to both publishers.

My dedicatees are three of the best people. Something of our shared genetic material has gone into this book. All three are wonderfully and totally themselves – but, though I took down the words here, I hope they see that we made it together. My thanks and my love.

T. D.
Bristol, 2018

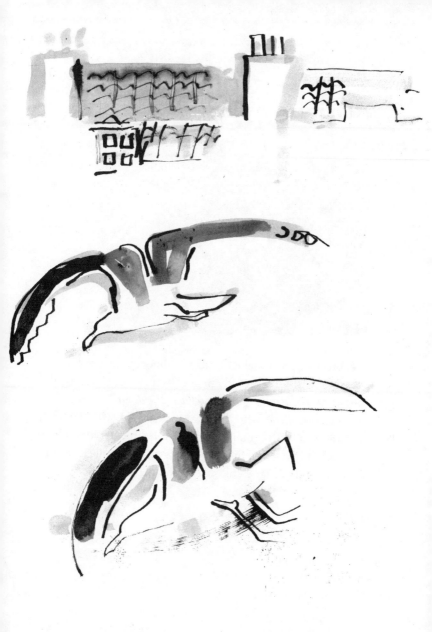

30518 - dormos windows dichension . no border units .

About the Author

Tim Dee has watched birds for almost all of his life and has written about them for twenty years. He is the author of *The Running Sky* and *Four Fields* and the editor of *The Poetry of Birds* with Simon Armitage and *Ground Work*. He was also a BBC radio producer for three decades. He is married and lives with his wife in Bristol, the Cambridgeshire fens and in the Cape Peninsula of South Africa.